Situating Open Data

Global Trends in Local Contexts

Edited by
Danny Lämmerhirt, Ana Brandusescu,
Natalia Domagala & Patrick Enaholo

AFRICAN
MINDS

Published in 2020 by African Minds
4 Eccleston Place, Somerset West 7130, Cape Town, South Africa
info@africanminds.org.za
www.africanminds.org.za

ISBN Paper 978-1-928502-12-8
ISBN eBook 978-1-928502-13 5
ISBN ePub 978-1-928502-14-2

Orders:
African Minds
4 Eccleston Place, Somerset West 7130, Cape Town, South Africa
info@africanminds.org.za
www.africanminds.org.za

For orders from outside South Africa:
African Books Collective
PO Box 721, Oxford OX1 9EN, UK
orders@africanbookscollective.com
www.africanbookscollective.com

Contents

About this book

The chapters in this edited volume represent contributions to the third Open Data Research Symposium (ODRS) held on 25 September 2018 in Buenos Aires. ODRS 2018 was chaired by Stefaan Verhulst (the GovLab, NYU Tandon School) and François van Schalkwyk (Centre for Research on Evaluation, Science and Technology, Stellenbosch University) with the invaluable support of the organising committee comprised of Emmy Chirchir (Münster University), Michael Cañares (Step Up Consultants), Tim Davies (Practical Participation), Gustavo Magalhães (University of Austin Texas and Portugal CoLab), Michelle McLeod (University of the West Indies), Silvana Fumega (ILDA) and Johanna Walker (University of Southampton). We would like to acknowledge the support of GovLab and Open Data for Development (OD4D) which helped organise the third edition of ODRS.

Acknowledgements

We would like to thank the following expert reviewers who were instrumental in the process of editing this book (in alphabetical order): Ingrid Brudvig (Web Foundation), Nkechi Cocker (Code for Africa), Eva Constantaras (Independent), Richard Heeks (University of Manchester), Glenn Mail (Web Foundation), Paul Mungai (Fairwork Foundation), Paul Plantinga (Human Sciences Research Council), Mor Rubinstein (360Giving), Giuseppe Sollazzo (Department for Transport), Johannes Tonn (Global Integrity) and Katherine Wikrent (Open Contracting Partnership).

About the editors

Danny Lämmerhirt is an ethnographer of digital data with an interest in public participation and critical data practices. Currently he is a PhD candidate at the University of Siegen, Germany. His dissertation examines how health data sharing platforms use data donations to turn personal data into 'valuable' resources and how notions of 'value' are constructed and contested. Danny was research lead at the Open Knowledge Foundation and assistant researcher with the Fraunhofer Society and at the University of Amsterdam.

Ana Brandusescu is an independent researcher, advisor and facilitator. She is the resident Professor of Practice for 2019–2020 at McGill University's Centre for Interdisciplinary Research on Montreal (CIRM) and an OpenNorth Fellow where she will design and implement a research agenda on AI in cities and its transformative effects on institutions. Ana previously led research and policy projects at the Web Foundation. She is on the advisory board of Learning from Small Cities.

Natalia Domagala leads on data ethics policy at the Cabinet Office, Government Digital Service in the UK. She previously advised on open government and open data policies for the Department for Digital, Culture, Media and Sport in the UK and implemented open data challenges for 360Giving. She has research experience in anthropology, gender, civic tech, and economic development. Natalia received her MSc in Local Economic Development from the London School of Economics and Political Science and her BA in Anthropology and Media from Goldsmiths University.

Patrick Enaholo holds a doctoral degree in media and communication from the University of Leeds, UK. His research interests include digital/social media and open data with particular focus on their cultural significance in society. He is currently a member of faculty at the Pan-Atlantic University in Lagos, Nigeria, where he also heads the Open Data Research Centre, a research unit focusing on the impact of data on development in developing contexts.

Introduction

The Open Data Research Symposium (ODRS) is a bi-annual gathering designed to provide a space for researchers working on open data to reflect critically on their findings and to apply and advance theories that explain the dynamics of open data as a socially constructed phenomenon and practice. ODRS is intended to be a safe space for debate in the presence of demands for fast results and proofs of impact – although this is not to suggest that researchers should be immune to considerations of relevance and transfer. Therefore, ODRS is usually organised alongside the International Open Data Conference (IODC) as a way for researchers to present their latest work, learn about other projects in the open data space and brainstorm new ideas. The symposium inspired the emergence of the Open Data Research Network with an aim to keep the momentum going in the ODRS community by sharing ongoing research.

For the 2018 edition of the symposium, the organisers received a total of 30 extended abstracts of which 12 were accepted for presentation in Buenos Aires. Selection was based on single-blind review of each abstract by at least two independent experts. Authors were required to submit full papers of their abstracts one week ahead of the symposium, and these papers were shared with those who had registered for the symposium. After the symposium, accepted authors were invited to revise and submit full papers for consideration in this edited volume. Papers underwent double-blind review by at least two peers, and authors were required to revise their papers before being accepted for publication.

Situating open data in critical data studies

The ODRS 2018 itself explored how open government data research should position itself within broader debates on data. Some attendees were in favour of including other types of open data beyond government data. Attendees also reflected on the fact that much open data research focuses narrowly on data from public institutions and governments, excluding research on open access and open science for instance. Others suggested the need to revive engagement with topics such as data ownership, the politics of making data public, or the relationship between open data, data rights and data justice. The role of open data for corporate actors, automated systems, artificial intelligence, and other emerging trends was also highlighted.

This suggests shifting viewpoints on open data initiatives from what they were some years ago. Instead of deploying and implementing open data programmes in social contexts, or seeking to understand their benefits and impacts, the value proposition behind these programmes should be interrogated, in order to zoom into the politics of the data. Data politics examines the influence data has on how humans relate to one another, who benefits from these relations, how value is created, and for whom (Ruppert et al. 2017). It offers an alternative to framing data as a resource to be unlocked, pooled, tamed, or unleashed, since it considers data as an essential element that is embedded in histories, social orders, situated ways of knowing, and a wider political economy.

This book provides empirical accounts to understand the situatedness of open data along the following themes: 1) open data practices; 2) the local implementation of global trends; and 3) open data ecosystems. Many chapters in this volume simultaneously address several of these themes. The thematic grouping of chapters is an attempt to foreground salient questions for open data research.

In addition, the book covers country-specific, localised applications of open data with a few chapters explicitly focusing on how open government data initiatives unfold within different socio-political contexts. The geographical scope of the contributions spans four continents, providing insights on open data practices in Europe (Kosovo, Belgium, United Kingdom), Africa (Nigeria, Tanzania), Asia (Indonesia, the Philippines), and Latin America (Paraguay, Brazil).

Open data practices

Several chapters discuss how actors practice open data. Each of them points – more or less explicitly – to the dual role data literacy plays. It may at once be a resource to upskill people and work with data and a form of critical practice that reflects on the historical functions of data within public institutions and the public sphere. Several chapters explicitly make a case for boosting data literacy as a matter of upskilling and point out resource and other constraints hindering the take up of data literacy.

Perhaps less explicitly, but not less importantly, several chapters critically examine how people practise data in different social settings – from health data reporting to data journalism – to show how data practice and the functioning of institutions are mutually related. Regarded in this way, data literacy becomes a form of critique that understands data not only as a product of institutions but as an active element shaping how institutions administer the world.

In Chapter 1: 'What technology and open data can do for women in Kosovo: A critical assessment of the potential of ICT skills programmes and open data to empower women in the ICT sector in Kosovo', Natalia Domagala critically assesses how open data is used in ICT grassroots programmes created for young women to gain ICT skills and work with open data, and whether this contributes to their empowerment. Domagala answers pertinent questions such as: What are the sources of disempowerment of women in the ICT sector in Kosovo? What are the strengths of ICT skills programmes and open data for the empowerment of women? What are their limitations? Through key informant interviews,

Domagala identifies that these programmes can counter the negative effect of socio-cultural norms, which end up discouraging women from undertaking employment in the ICT sector. These programmes do so by providing highly demanded skills, support networks and role models. Yet, ICT skills building programmes can exclude women who lack access to the internet, provide little practical employment assistance and fall short on strategic ways to strengthen political engagement even though they are often touted as championing civic engagement. The chapter provides practical insights for policymakers on how to improve the effectiveness of ICT skills programmes and their ability to create broader social benefits by using open data.

In Chapter 2: 'Journalists and the intermediation of open data: A Nigerian perspective', Patrick Enaholo and Doyinsola Dina examine the work of open data intermediation by journalists in Nigeria. The chapter discusses how journalists, as intermediaries of government data, engage with end-users in the open data ecosystem. The chapter builds on the competency framework proposed in Van Schalkwyk et al. (2016) to understand the role of journalists as intermediaries of open data. Using a mixed method approach, the authors determine that, within the specific context they operate, technical as well as creative competencies are key to ensuring the effectiveness of journalists' role in the open data ecosystem. Although the study also identifies the lack of technical skills and competencies as the primary challenge for journalists in the Nigerian context, they show how these are compensated for by other intermediaries in the open data value chain such as civil society and civic tech groups.

In Chapter 3: 'Using open data for public services', Miranda Marcus, Ed Parkes, Therese Karger-Lerchl, Jack Hardinges and Roza Vasileva explore the use of open data in public service delivery in a collaborative manner to solve problems and support innovation in the UK public sector. The chapter distinguishes three patterns of open data in public sector institutions and their effects on delivering public services: (1) open data for increasing access to public services; (2) open data for more efficient public service delivery; and (3) open data for policy development. The research has been conducted through a combination of interviews, desk research, and visualisation techniques. Due to the complexity and non-linear process of public service delivery, the authors used a rich picturing method to portray public services as ecosystems. The chapter concludes with comparisons across the three patterns of open data use in public services, looking at the following aspects: organisational collaboration, technology infrastructure, digital skills and literacy, foundational data infrastructure, open standards for data, senior-level championing, intermediaries, problem focus, open innovation and peer networks.

Local implementation of global trends

In Chapter 4: 'Localising global commitments: Open data in sub-national contexts in Indonesia and the Philippines', Michael Cañares analyses the role and scope of local government in implementing the Open Government Partnership commitments. The chapter develops case studies of early open data adopters in Bohol province in the Philippines and the city of Banda Aceh in Indonesia. The cases describe how local open data initiatives are shaped by political leadership, civil society, and data literacy. It then outlines challenges and opportunities for implementing open data initiatives. The chapter concludes with recommendations for open data practitioners in local governments to: 1) engage in local consultations; 2) ensure a legal basis for open data initiatives; 3) strengthen citizen engagement; 4) and invest in local capacity for data suppliers and users.

In Chapter 5: 'Closing the gaps in open data implementation at sub-national government level in Indonesia', Ilham Cendekia Srimarga and Markus Christian trace the implementation of an open data project in a sub-national context. The authors discuss their involvement in the transition of Bojonegoro district government in Indonesia from one stage of aggregated data to one in which disaggregated data is employed in order to facilitate higher levels of public participation with open data. The challenges encountered during the first stage included the obsolescence of the data and the difficulty of processing due to its closed format which fell short of expectations and led to a predominantly low response from citizens. The second stage involved civil society organisations and other stakeholders to provide analyses and recommendations that eventually engendered a 'data revolution' concept. The authors examine how data provisioning technology (such as the format and representation of open data) and the context in which this happens (such as the low rate of adoption of open data by stakeholders) can shape the eventual outcomes of such implementations.

In Chapter 6: 'The cost of late payments in public procurement', Juan Pane, Camila Salazar, and Julio Paciello examine how granular open data on public procurement payment processes can be used to contribute to cost savings for public administration. Based on public institutional data from Paraguay, the authors assess the duration of public procurement payments to estimate the cost of payment process delays and how they affect suppliers. In turn, the authors determined the variables that contribute to these delays using descriptive analysis and financial cost estimation to calculate the cost of late payments on suppliers, and modelled the duration of payments using survival analysis to identify which variables have a role in delaying payments. When properly implemented and analysed, the study found that open data can positively impact the public procurement cost saving process. The study finds that the duration from the moment an invoice is issued to when the payment occurs can be of approximately 55 days on average for each payment as opposed to international practice of a 30-day average. Between 2011 and 2017, late payments accumulated to USD 81.07 million. For cost saving efforts to be successful, the authors recommend the Paraguay public administration to analyse the payment process and apply an appropriate corrective normative framework.

In Chapter 10: 'From theory to practice: Open government data, accountability and service delivery', Michael Jelenic interrogates the extent to which the use of open government data is associated with higher levels of accountability and improved service delivery in developing countries. Constructing a unique dataset that operationalises open government data, government accountability, service delivery, as well as other intervening and control variables, the research adopts secondary data from 25 countries in sub-Saharan Africa to find a number of significant associations between open government data, accountability, and service delivery. However, the findings suggest differentiated effects of open government data across the health and education sectors, as well as with respect to service provision and service delivery outcomes. Although this early research has limitations and does not attempt to establish a purely causal relationship between the variables, it provides initial empirical support for claims about the efficacy of open government data for improving accountability and service delivery.

Open data ecosystems

Several chapters attend to the constituent parts of open data ecosystems to understand how these can be infrastructured for broader socio-economic benefits. The chapters reflect a recent academic debate among platform studies and infrastructure studies scholars who argue that there is a hybridisation of concepts usually considered apart from another. Platforms that are usually considered as programmable, modular systems become infrastructural, meaning engrained in daily lives and serving critical functions. Infrastructures which are usually considered as learned practice, critical for daily lives, and visible only during breakdown become increasingly 'platformised', meaning splintered across service providers (Plantin et al. 2016).

François van Schalkwyk's contribution to this book adds to this discussion, reflecting on how an open data platform could shape data flows in existing information infrastructures. In Chapter 7: 'Connecting flows and places: Flows of (open) data to, from and within hyperlocal communities in Tanzania', he discusses how health-related data travels in a data ecosystem between public institutions and civic actors. Building on theories of data ecosystems, the chapter maps out how infrastructural elements shape 'blockages, breaks, switchers and connectors of data flows'. These infrastructural elements include databases and how they connect to data exchange, reporting procedures to digitise and share data, as well as trained social conventions and (dis-)incentives to report and share data across local, regional, and national scales. Using the method of tracing data flows, the chapter draws attention to how data flows can have centralising tendencies, with many reporting routines steered towards higher-level administrations, instead of enabling data flows towards civic communities. The chapter provides fresh thinking not only on how open data platforms can integrate in information systems, but also how data flows are shaped by organisational interests, and how they create inherent conflicting interests, exclusion, as well as power dynamics.

Other contributions approach open data ecosystems from the viewpoint of platform architectures to understand how the design of platform elements (such as APIs) interact with the performance of application ecosystems and user satisfaction. In Chapter 8, Julián Rojas, Bert Marcelis, Eveline Vlassenroot, Pieter Colpaert, Matthias van Compernolle and Ruben Verborgh discuss 'Decentralised open data publishing for the public transport route planning ecosystem'. The chapter studies the implementation of a decentralised open data publishing strategy in the public transport route planning ecosystem. Comparing two data architectures – REST APIs and decentralised linked data – the chapter discusses how each enables users to withdraw, compile, and calculate different types of data. The chapter provides insights on how platform elements influence the performance of network components, including apps and mobile phones. In addition, the chapter provokes thinking on how the platform architecture relates to privacy-by-design, technical performance, and user-perceived satisfaction with traffic routing apps.

In Chapter 9: 'Building a framework for analysis of factors to creation and growth of an open data ecosystem', Edson Carlos Germano, Nicolau Reinhard and Violeta Sun propose a framework for analysing and exploring the factors that can influence the economic and social sustainability of Open Government Data (OGD) ecosystems comprising producers, intermediaries, and users who maintain relationships with each other by creating a flow of information, services, products and financial resources. The authors set out to examine the information flow throughout the OGD value chain with a focus on data reuse in emerging countries. Based on an in-depth literature review, they propose five characteristics for the analysis and understanding of OGD ecosystems. The proposed framework considers rules and governance aspects defining the relations between ecosystems' data producers and final users of the products and services, focusing on informal membership rules in the ecosystem. The authors argue that with such a framework – which is currently being adopted for the Brazilian federal government OGD ecosystem – it will be possible to analyse and understand the governance necessary for the construction and sustainability of OGD ecosystems. This, to the authors, can help society obtain greater benefit from using OGD analysis to understand how government resources are used more efficiently.

We hope that this book sparks a critical reflection on the way open data is practised, implemented locally, and its strategic role in ecosystems. The volume departs from framings of data as a resource, and on impact as an end-goal. Rather, the chapters focus on the process and how open data influences social relations, shapes histories and power structures affecting information systems.

The editors
October 2019

REFERENCES

Plantin J-C, Lagoze C, Edwards PN & Sandvig C (2016) Infrastructure studies meet platform studies in the age of Google and Facebook. *New Media & Society* 20(1): 293-310. https://doi.org/10.1177/1461444816661553

Ruppert E, Isin E & Bigo D (2017) Data politics. *Big Data & Society* 4(2): 1-7. https://doi.org/10.1177/2053951717717749

Van Schalkwyk F, Cañares M, Chattapadhyay S & Andrason A (2016) Open data intermediaries in developing countries. *Journal of Community Informatics* 12(2): 9–25. http://ci-journal.net/index.php/ciej/article/view/1146

1.

What technology and open data can do for women in Kosovo: A critical assessment of the potential of ICT skills programmes and open data to empower women in the ICT sector in Kosovo

Natalia Domagala

Information and communication technologies (ICTs) are a powerful force driving transformations of social, political, and economic processes globally. For many scholars (Hafkin & Taggart 2001; Hafkin et al. 2003; Huyer 2006; Jain 2015), the emergence and popularisation of these new tools present an opportunity to promote sustainable growth, especially in developing countries.

However, many recognise the risks that unequal implementation of new technologies presents, particularly for women – in this research, noted as cisgender and self-identified women – and people with limited access to them. Despite significant progress in the gender equality agenda, if the benefits from ICTs are determined by existing power relations, these technologies and their implementation will not be gender-neutral (Nath 2001; Gurstein 2011; Hilbert 2011). Generally, women around the world have less access to and control over ICTs and thus remain marginalised within economic systems (Ng & Mitter 2005; Oliver 2017; Huyer 2016). The concern is that this digital divide will deepen if existing divisions are not addressed. Therefore, it is critical to ensure that women have equal access to ICTs and the right skills to fully harness the opportunities that the knowledge economy presents (Hafkin & Taggart 2001; Gurumurthy 2004). This raises significant concerns of how to increase the number of women interested in technology and their ability to shape its future, as well as how to leverage their position within the industry. Creating grassroots, women-only, ICT skills programmes has been identified as a force for empowerment that could lead to positive economic changes in the long run. A growing body of research analyses the potential benefits of such training and its ability to support various aspects of the empowerment of women (Jorge 2002; Gurumurthy 2004; Jain 2015; Lee 2004).

This chapter aims to address these issues by examining the potential of ICT skills trainings and open data, a tool commonly used during such trainings, to empower women within (but not exclusively to) the male-dominated ICT industry in Kosovo. Given extremely low female labour and political participation in the country, Vidovic et al. (2017) propose that a combination of policy and cultural changes could break the existing links between various factors driving inequality. These factors include less schooling, early marriage, and entrenched patriarchal family and community attitudes about the fulfilment of traditional roles. Digital skills could emerge as a progressive force by simultaneously empowering women and creating high-skilled human capital that would contribute to economic growth in the country.

While many separate aspects of empowerment are well-documented in the literature, a holistic approach exploring its various dimensions is less common. This chapter merges definitions of empowerment from feminist theory and economics (Kabeer 1999; Cornwall 2006; Balitwala 1993; Lennie 2002) with the findings of ICT skills literature to explore the effectiveness of ICT programmes in empowering women in four spheres: social, political, psychological, and technological. The organisations examined in this study are two grassroots civic technology initiatives: (a) Open Data Kosovo (ODK), focused on open data, governance, transparency, and accountability; and (b) Girls Coding Kosova (GCK), working towards a free-of-gender-bias ICT sector in Kosovo and the region. Data was collected through 20 semi-structured interviews with the organisers and participants of the programmes. The main question the research seeks to answer is as follows:

To what extent do grassroots ICT skills programmes contribute to the empowerment of women in the ICT sector in Kosovo?

To fully address this research question, three subordinate questions will be examined:

1. What are the sources of disempowerment of women in the ICT sector in Kosovo?
2. What are the strengths of ICT skills programmes and open data for the empowerment of women?
3. What are their limitations?

The chapter confirms existing literature findings, in that the main source of disempowerment derives from persistent social norms, followed by discrimination, discouragement of women from pursuing employment in the sector, and scepticism of their technical capabilities. ICT programmes help to alleviate these obstacles by equipping participants with relevant skills, building support networks, improving their confidence, and engaging them with open data – an

ICT tool commonly used to encourage active citizenship. Nevertheless, the study exposes some major limitations of the programmes such as their exclusivity, a lack of practical employment assistance, and, despite progress in civic empowerment, a shortfall in strengthening women's political engagement.

The study is relevant and timely considering the increase in popularity of ICT skills trainings and open data as developmental tools targeting women around the world. This chapter contributes to research on ICT-based empowerment by merging existing models to assess the effectiveness of initiatives aimed at improving the position of women within their societies. Not only does it address a gap in the literature on holistic approaches to empowerment in the ICT context by combining feminist theory with ICT studies, but it also provides a set of guidelines for ICT policymakers on what works in the implementation of skills programmes. Moreover, it situates open data within a broader context of ICT-enabled empowerment of women.

The chapter proceeds as follows: section 2 establishes the conceptual framework by reviewing the key literature on empowerment and ICT skills. Section 3 focuses on the geographical context of Kosovo. Section 4 describes the methodology. Section 5 presents and analyses the findings. The final section concludes and outlines broader policy implications of the research.

Literature review and conceptual framework

Empowerment

It is commonly recognised that empowering women through education and training will increase overall economic development (Wilson 2015; Prügl 2017). A large body of gender and development literature attempts to apply the notion of empowerment to developmental initiatives targeting women, such as microcredit or entrepreneurship programmes (Batliwala 1993; Lennie 2002; Sanyal 2009; Cornwall 2016; Kabeer 2011; Milazzo & Goldstein 2017). However, few studies establish comprehensive guidelines for a qualitative measurement of empowerment, specifically in relation to projects focused on ICT. This section aims to analyse and synthesise the existing conceptualisations of empowerment into a framework that can be applied to the assessment of ICT skills programmes.

The notion of empowerment has been equally praised and criticised for its fuzziness (Batliwala 1993; Kabeer 1999). The term can refer both to the power given to an individual by an external source, and to the process whereby an individual gains control of their life. This seemingly contradictory definition has sparked an academic debate on whether empowerment is a passive or an active process and how it can be achieved. Previous studies (Batliwala 1993; Kabeer 1999) indicate that changes in the perception of the world can trigger the will to collectively re-establish power structures. Therefore, conceptualising 'power'

3

is critical. In her understanding of power as executing control, Batliwala (1993) deconstructs the composition of power relations. For Batliwala, the central aspects are control over resources (physical, human, intellectual, financial, and the self), as well as control over ideology (beliefs, values and attitudes). Hence, if power equals control, empowerment becomes the process of gaining control of the aspects listed above. Kabeer's (2017: 651) interpretation extends this definition of empowerment as shifting control over resources and ideology to the capacity of individuals to make meaningful choices about their lives by those who have been denied this ability to make choices. This enhances Batliwala's understanding of the concept by placing more emphasis on the relational aspects of the process. Thus, for Kabeer (2017: 651), the key is for these choices to happen in a manner that actively challenges and changes the structures of inequality in a society.

What becomes apparent from these definitions is that empowerment is relational and context dependent. Cornwall (2016: 344) argues that it has no fixed start or end points as it constantly evolves – empowerment can be temporary, and what empowers one woman, might not empower another. With these factors in mind, a suitable definition of the term in the context of assessing the empowering capabilities of ICT skills programmes should convey the complex interactions between individuals, socio-cultural norms, and emerging technologies. However, a limitation of the definitions from the previous studies (Batliwala 1993; Kabeer 2017) is their lack of consideration of the process during which various resources appear in a society. Given that ICT is a rapid, disruptive force, it is crucial to perceive it as distinct from pre-existing endowments. ICTs differ from other resources in their ability to challenge existing power structures by providing connectivity and access to information (Pollock 2018). Therefore, a framework analysing the empowering effect of ICT skills programmes should recognise technology as a separate factor.

A model of empowerment designed by Friedmann (1992) and adapted by Lennie (2002) presents empowerment as a multidimensional process, engaging the individual, the community, and the relationships between them by distinguishing four categories: a) social (participation in organisations, social capital); b) political (access to the decision-making processes, power of voice, vote and collective action); c) technological (developing ICT skills and technological competencies, being able to find employment in the field); and d) psychological (confidence, individual sense of potency) (Friedmann 1992; Lennie 2002). Here, technology is placed within a broader socio-political context and the interdependency between different types of empowerment is captured; for instance, political empowerment usually requires prior social empowerment. What this model lacks is tracking the process of change. Thus, it could be improved by incorporating features of the framework (see Figure 1) collated by Cornwall (2016) and deployed in an ICT-related study by Bailur et al. (2018). By synthesising the main factors influencing the lives of women and highlighting interactions between them, their framework clearly outlines the process and changes necessary for achieving greater empowerment. It thus illustrates that sustained empowerment requires a change in human consciousness as well as in formal laws and cultural norms.

Merging those two models (Figure 2) provides an understanding of both the components and the processes required to stimulate change and will be used in this study. This will enable a holistic understanding of how to achieve empowerment in these four distinct spheres.

Figure 1. Cornwall's conceptualisation of empowerment

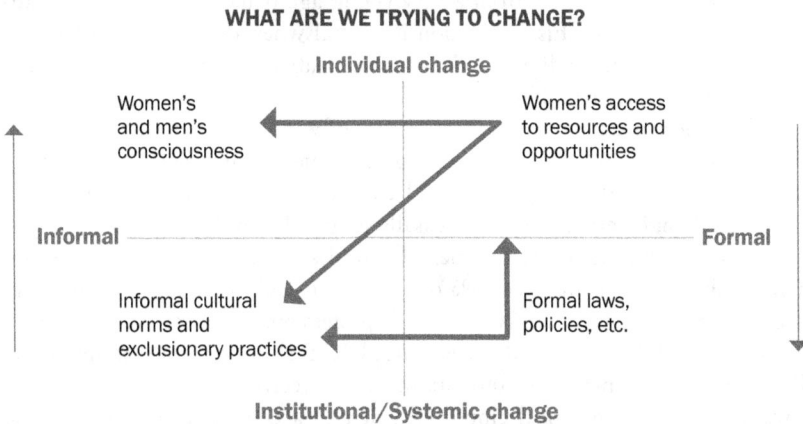

Source: Cornwall (2016: 346)

Figure 2. Multiple components and processes of change

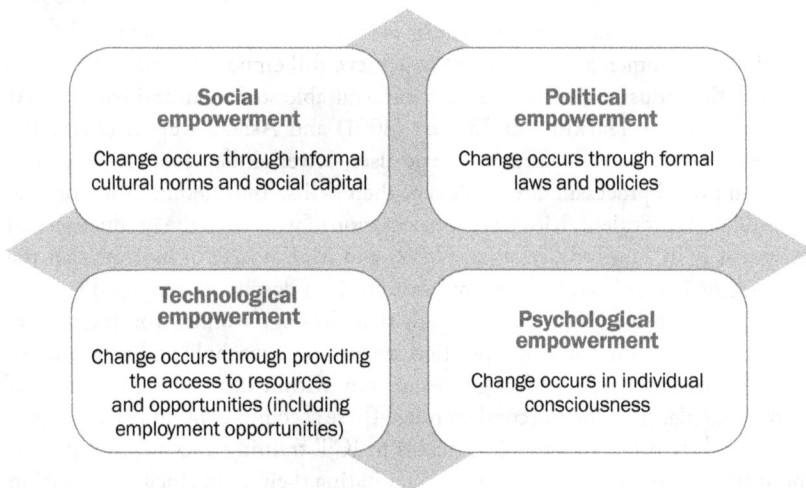

Technological empowerment

In recent years, there has been an increasing amount of literature describing ICT skills as a pathway to prosperity and emphasising the necessity to become fluent in digital technologies (European Commission 2016; Ananiadou & Claro 2009; Prensky 2001). The emergence of new technologies transforms the nature of work and enables citizens to participate in social and civic life in previously unknown ways. Thus, possessing at minimum a basic level of ICT literacy is the new norm. The younger generation of 'digital natives' are born into the world with a natural advantage over their older counterparts, as well as those who were unable to access ICT from a very young age (what Prensky (2001) calls 'digital immigrants'). This distinction is crucial when designing curricula for ICT skills trainings, as digital abilities of the 'natives' are more advanced, and their learning methods differ.

While there is no universal classification of what constitutes as ICT skills, throughout this chapter the term will be used to refer to advanced skills including (but not limited to) coding, front and back end software development, web design, and engineering. The digital skills supply-demand gap has been widely recognised: according to the European Commission (2016), 90% of future jobs will require digital skills, yet 44% of Europeans lack basic digital education. Considering that less than 20% of ICT professionals are female (European Commission 2016), the need for advanced ICT skills trainings is immense. Thus, designated female-only programmes are perceived as one of the pathways towards achieving technical empowerment and access to new opportunities. As discussed by Nath (2006), Gurumurthy (2004) and Jorge (2002), learning the skills essential to fully utilise the benefits that ICTs offer has the potential to empower women by enhancing their capabilities to overcome the barriers preventing them from fully participating in economic and social processes.

However, what these studies dismiss is that providing access and ICT skills trainings for women are insufficient to achieve full empowerment if the norms shaping the industry are aligned with unfavourable socio-cultural patterns. As demonstrated by Hafkin and Taggart (2001) and Ashcraft et al. (2016: 19), women tend to be concentrated in end-user, lower skilled jobs, such as data entry and word processing, regardless of their skills. Thus, feminisation of low-skilled jobs is prevalent, with low representation of women in the production and design of ICT. Similarly, Morgan (2012) and McKay (2006) indicate that the ICT sector is regulated by gendered assumptions despite its supposed gender-neutrality. McKay's (2006) study reveals that although women working in low-skilled ICT jobs improve their position within the household and gain relative financial autonomy, those changes occur only within the existing divisions of labour that placed them in secondary roles. These accounts contradict the popular assumption that merely providing access to ICT training and infrastructure is the answer to empowering women and facilitating their digital inclusion. Hafkin and Taggart's (2001) argument that women still remain the passive consumers of ICT, which discredits them from undertaking leadership roles, demonstrates that the acquisition of specialist skills is inhibited by social norms, and that women's positions in the industry are diminished by default. Considering that the knowledge economy relies heavily on ICT and will do so even more in the future, if women are excluded from the design and implementation of new tools, programmes and systems, unequal power relations will prevail despite the potential to renegotiate them in light of systemic changes of the technology-based order. As a potential solution, Huyer (2006: 27) recommends that women create and develop technology and the content it carries. Therefore, the emphasis should be on improving women's position within the industry by teaching them more advanced expertise and creating routes into high-skilled employment.

Social and psychological empowerment

A key aspect of female-only ICT skills programmes is that they have the potential to support women in broadening the scope of their actions by providing them with social capital. Across the literature, the term refers to the extent, nature and quality of social ties that individuals or communities can mobilise (Zinnbauer 2007: 16) through the use of resources such as trust, information, social norms (Coleman 1990; Lin 2001; Putnam et al. 1993), that facilitate cooperation within or among groups (OECD 2007). So far, few studies have covered the role of social capital in ICT trainings, and this chapter aims to address this gap. However, gender and development scholars have thoroughly analysed the formation and influence of social capital within women-only development programmes (e.g. microfinance), concluding that in most cases such programmes foster social cooperation and enhance social networks (Sanyal 2009; Pitt et al. 2006; Maclean 1999; Karim 2008), with factors such as economic ties and regularity of meetings facilitating a shared sense of community and the capacity to undertake collective action. Therefore, it could be argued that such trainings create spaces of interaction and strong reciprocity due to their basis in the altruistic sharing of skills and experiences (Bowles 2002). However, as suggested by Coleman (1990), social capital emerges as a by-product, rather than as the main aim of these initiatives, indicating that a potential limitation of the trainings might be an insufficient emphasis on community-building activities.

One implication for ICT skills programmes is that through facilitating bonding (connections between similar people) and bridging (connections beyond a shared sense of identity) they can help assemble communities of women willing to support each other outside of the trainings themselves – for instance when looking for jobs. A large body of research indicates that social networks are vital in the process of finding employment, mainly through referrals and reducing deficits of information in the labour market (Dustmann et al. 2016; Zinnbauer 2007; Sabatini 2009; Stoloff et al. 1999). Therefore, having an extensive network might be crucial for women when searching for employment in in the male-dominated ICT sector (Franklin et al. 2005), where discrimination and scepticism may hinder their abilities to find jobs through the formal routes. Several feminist studies have identified the exclusionary and discriminatory aspect of male social capital working to maintain the status quo (ibid.) in the traditionally male-dominated sector (Henwood 2000; Cockburn 1983). While female-only networks facilitate women finding jobs, exclusive, better-established male networks could undermine this effort. This is supported by Fukuyama's (1999) assumption that social capital produces negative externalities when group solidarity is created at a price of hostility. However, as demonstrated by Storper (2005), bonding enhances the potential for autonomy and builds people's capacities, while bridging limits the potential opportunism of hostile actors. This proves that in order to contradict male domination in the ICT industry, maintaining strong social ties between women is essential.

Moreover, ICT trainings facilitate bridging by bringing women from the same location but different ethnic groups together to work towards shared goals. There is a consensus in the literature that local ICT initiatives can help to create trust and shape serendipitous social networks (Gaved & Anderson 2006; Williams 2005; Blanchard & Horan 2000), for instance through regular face-to-face meetings or messaging groups and forums.

Participation in a community, understood as a group of people with shared identities, expectations, and interests (Storper 2005: 34), stimulates psychological gains on an individual level (Zinnbauer 2007; Mishra 2016). For Lennie (2002) and Mishra (2016), psychological empowerment encompasses rises in self-confidence and self-esteem, increases in motivation, enthusiasm and self-value, well-being and feelings of belonging that promote freedom and a willingness to express oneself. This rise in psychological well-being is perceived as crucial to succeeding in the workplace (Mishra 2016; Zinnbauer 2007; Valarmathi & Hepsipa 2014), perhaps because it increases people's own beliefs in their abilities to achieve their goals and control their environment (Zimmerman 2000). This view is supported by Oladipo (2009), for whom psychologically empowered people – through a shift in their attitude, cognition and behaviour – have the capacity to stimulate positive changes on a social scale. Finally, some of the potential sources of psychological disempowerment defined in the literature include fear, insecurity, risk, lack of self-esteem and self-confidence, and a fear of failure (Mishra 2016; Valarmathi & Hepsipa 2014). Nevertheless, little has been said about the issue of psychological empowerment as it relates to ICT skills programmes specifically.

Political and civic empowerment

In their seminal study on social capital and democracy, Putnam et al. (1993) argue that social networks and norms are key to building and maintaining a successful democracy. If ICT skills trainings and the knowledge networking they create are a source of strong female social capital, they could be deployed to increase women's political empowerment.

Scholars including Nath (2006) and Hafkin et al. (2003) emphasise that ICT trainings can influence political participation in two ways: first, by providing more information that was not previously available, and second, by bringing together larger groups of women mobilised around mutual concerns. They argue that coalitions formed during the shared experience of ICT skills learning have the potential to strengthen decision-making as well as local and national democratic processes. Despite this valuable insight, the studies offer no recommendation for how to interest ICT communities in political and civic engagement. Therefore, the challenge lies in designing and implementing ICT skills programmes that spark and encourage interest in these issues. The link between ICT skills and politics is not straightforward. However, the field of civic technology can serve as a method to enhance citizen participation in governance processes while simultaneously working to improve digital infrastructure and digital service delivery in the public sector (Peixoto et al. 2017). A crucial tool with the potential to connect the two fields, as proposed by internet and governance scholars (Ruijer et al. 2017; Pollock 2018), is open data. Understood as data that can be freely used, shared and built-on by anyone, anywhere, for any purpose (James 2013), it is one of the key concepts in civic technology. Open data is seen as an effective tool to counter corruption because it allows citizens to monitor government spending and view contract data. Hence, if shared and used comprehensively, it enables citizens to hold their governments accountable (Boland & Coleman 2008; Davies 2010).

However, Noveck (2009) is sceptical of tech-enabled civic communities. She challenges their transformative powers by pointing out the fact that despite bringing people together, civic programmes are disconnected from governmental practices themselves, and thus fail to change the ways in which institutions obtain and use information (Noveck 2009: 31). This valid criticism challenges us to rework previous models of citizen participation to incorporate the modernisation of governance systems where citizens can actively work with data and available information. For open data to be truly empowering, public participation and social engagement should be normalised and citizens should be involved in designing responses to public needs (Ubaldi 2013). This creates benefits for the governments – crowdsourcing of services and content can improve the governments' effectiveness and enhance policy-making through deploying the talent from outside of the public sector to use open government data. It is this ability of third parties to participate in and shape the use of open data that makes it a transformative and empowering tool. As concluded by Noveck (2012), open data enables those with technical know-how to create tools, models, and visualisations that provide innovative insights and bring fresh perspectives and resources. One example of this is the Code for America programme, where software developers design digital solutions for the needs of their local governments.

The use of open government data is a socially situated process that operates within broader social arrangements, power dynamics and market conditions (Davies 2012; Felten 2009), hence the effectiveness of open data initiatives can be determined by these non-technical factors. Therefore, creating a culture of regular open data use and citizen engagement is essential for the efficient use of the data to support democratic processes and increase civic participation. Similarly, although such programmes increase civic awareness and participation, perhaps with an eye towards more participatory democratic models, their potential for politically empowering women might be limited if the social constraints preventing women from thriving in the ICT industry are not addressed first.

Moreover, for the political benefits of ICT skills and open data to be reaped, a high level of skills (Gurnstein 2011) and education on the requirements and responsibilities of citizenship is required (Weare 2002). Previous studies on the political empowerment of women through the use of open data and ICT skills have had mixed results. In their report on the impact of open data on women in Africa, Brandusescu and Nwakanma (2018) identify cultural and social realities as hindering women to engage with data and participate in the technology sector. Furthermore, the authors point out that the lack of a strong open data research base and women's digital groups prevents open data from being used in its full capacity for political and civic empowerment (Brandusescu & Nwakanma 2018).

Therefore, although digital skills, civic technology and open data could instil democratising aspirations in citizens, educators, and political authorities, their potential to empower women specifically is debatable.

Kosovo: The youngest and the poorest

Kosovo is by no means an ordinary place. Due to its turbulent history and uncertain political status over the years, the country is now facing a distinct set of developmental challenges that this section aims to explore.

Kosovo is Europe's youngest country, both in terms of history and demographics (Curto & Simler 2017). It declared independence from Serbia on 17 February 2008, yet its international recognition remains partial. After a four-and-a-half-year period of supervision by the UN, Kosovars built their institutions from scratch (Curto & Simler 2017: 20). However, the legacy of post-Yugoslavian political culture and widespread informality posed a challenge to forming transparent, credible and efficient institutions, with corruption emerging as a persistent problem. The lack of a coherent social infrastructure such as the rule of law, property rights and an ineffective judicial system, alongside the country's disputed status, further complicate the political landscape. Therefore, the emergence of a strong civil society able to subject its governing bodies to greater pressure to increase transparency is crucial in shaping Kosovar democracy.

As aforementioned, demographically, Kosovo is the youngest country in Europe. Out of its total population of 1.8 million, around 38% of the population is younger than 19 (Curto & Simler 2017: 20). Although a young population is an asset and a potential resource for future prosperity, if not utilised properly it can lead to further destabilisation of the country. Currently, the potential of Kosovo's youth remains largely untapped: in 2015, nearly one-third of young Kosovars were neither in education or training and were unemployed. There is a substantial risk that the young population will become a 'demographic curse', leading to outward migration and brain drain, if the state fails to provide them with attractive future prospects. However, tough visa regulations and the fact that Kosovo remains the only territory in south-eastern Europe whose citizens require a visa to travel to the Schengen Area (Curto & Simler 2017: 108), constrain any current labour mobility.

Poverty is another major problem. Despite Kosovo's steady economic growth, the country is still the poorest in Europe (Diakonidze et al. 2016; Curto & Simler 2017). Kosovo also exhibits common characteristics of small states such as a narrow production base due to high input costs, a small internal market, tariff-dependent government revenues, and FDI and exports at a level insufficient to transform the economy.

The combination of all the factors listed above makes Kosovo a striking case where technology, digital skills, and open data could form a partial solution to issues of poverty, youth and female unemployment and weak governance. The ICT industry has been recognised as having the most potential for future development, and a number of government strategies made it a priority. However, inadequate human capital, poor quality of education and relatively low levels of digitalisation in the economy constrain this potential. (Ministry of Economic Development Kosovo 2013; Diakonidze et al. 2016; Shala & Grajcevci 2018).

This chapter argues that women are particularly affected by these problems. They are disadvantaged in the political, economic and social spheres, as well as underrepresented in public administration, where only one position in five is occupied by women (Curto & Simler 2017: 146). Furthermore, surveys of the ICT sector in Kosovo indicate that there are four times more men employed in the ICT sector than women. Additionally, 24% of the surveyed women admitted that they decided against working in ICT fields due to their perception as 'male professions', while 44% claimed that they lacked adequate support networks (Diakonidze et al. 2016). Lack of assistance from employers and gender prejudices constitute other significant hindrances.

Thus, there is a need to upskill the female workforce in Kosovo and to create support and employment networks that can compensate for their unfavourable position. There has been a consensus in the literature (Diakonidze et al. 2016;

11

Kelly et al. 2017; World Bank 2016) that the key to the future successful economic development of Kosovo is to transform latent talent into productive human capital, with young women being the major target group. Therefore, ICT skills programmes targeting young women could improve the competitive position of Kosovo in the global economy. Girls Coding Kosova and Open Data Kosovo are the main grassroots organisations tackling the challenges outlined above, and their work will now be discussed.

Girls Coding Kosova and Open Data Kosovo

Blerta Thaçi started Girls Coding Kosova (GCK) when she noticed the contrast between the number of women studying computer programming and women working in the IT industry. GCK started as informal gatherings of women working in technology who would discuss their experiences. The organisation aims to create a free-of-gender-bias IT sector in Kosovo by increasing the number of women interested in programming, providing opportunities for them to gain skills and experience, and exposing women to coding and software programming from an early age.

Open Data Kosovo (ODK) is a civic technology organisation promoting open data and governance. ODK focuses on transparency, accountability, publishing business and procurement data, raising awareness and teaching ICT skills. Blerta Thaçi was also working for ODK and eventually became their executive director. Now, the organisations share the office and work on most of their projects together.

GCK projects target young women (18–25 years old), primarily, but not exclusively, from the ICT sector, and cover the whole range of skills currently in demand in the market while promoting technology-motivated civic engagement and open data driven solutions. The portfolio of both organisations includes Tech4Policy,[1] a programme where female residents design digital tools to address problems in their municipalities; Code4Mitrovica,[2] a project bringing together Serbian and Albanian Kosovars to work on tech-based solutions for Mitrovica, a disputed city in the north of Kosovo; Techsperience and Techstitution, a programme through which the youth from the ICT community develop digital tools that improve the quality of work in Kosovo's local and central level institutions; as well as a number of reports and platforms, including the Open Government Data Platform, Open Business, and Walk Freely, an app for women to report sexual abuse and receive data with patterns of street harassment.

---

1   http://tech4policy.com/
2   https://www.facebook.com/GirlsCodingKosova/videos/850368838474744/

Methodology

One motivation behind choosing ODK and GCK as subjects of this study derives from Kosovo's socio-economic characteristics, which provide unique conditions under which to observe the effect that ICT skills might have on the lives of women. The effects of grassroots initiatives aiming to empower women can be captured more explicitly in Kosovo than in countries with higher levels of educational attainment, less unemployment, and fewer cultural constraints for women. Moreover, the attempt to increase female civic engagement through the popularisation of open data-driven initiatives makes ODK and GSK pioneering examples of a holistic approach to development, particularly important in the case of young democracies.

There are no standard indicators to measure gender-specific empowerment (Kabeer 1999; Lee 2004; Bailur et al. 2018). Therefore, each study on the topic requires its own bespoke conceptual framework. Although the elusiveness of the concept can be a hindrance, as mentioned by Kabeer (1999: 436), for many researchers, its value lies in its 'fuzziness' and in the lack of a clear, universal definition. Therefore, one of the most successful methods of studying empowerment is through individual experiences and narratives gathered qualitatively. Thus, semi-structured interviews and conversations enable a better understanding of the factors that contribute to the failures or successes of ICT skills programmes for the empowerment of women. Due to the definition of empowerment adapted to the needs of this study and the local context of Kosovo, quantitative measures would fail to convey the complexity of the socio-economic implications of ICT skills trainings and the changes they initiated in the lives of individual women. Moreover, semi-structured interviews have been adopted by similar studies analysing the possibilities and hindrances that ICT skills trainings present for women (Lee 2004; Masika & Bailur 2015; Hussain & Amin 2018).

The data essential to establish whether digital skills programmes have an empowering impact on the lives of women in Kosovo were gathered through in-depth, semi-structured face-to-face interviews with a group of 20 stakeholders. This consisted of nine participants and eleven organisers of the courses (five of the interviewed organisers are past participants), including the founders of ODK and GCK, and the executive director of both organisations. The majority of interviewees were women under 30. Due to the specifications of the research question, the initial research plan did not include any male participants. However, this was revisited while conducting fieldwork and eventually, three men were interviewed in order to understand their perception of female empowerment as male members of the organisation – the founder of ODK, a member of the technical team, and one of the mentors. 65% of total interviewees and 78.57% of all the participants had previous information technology (IT) backgrounds. The interviewees were selected to represent the variety of programmes organised or supported by GCK and ODK to help ensure that the findings were not limited to a one particular course and the process of acquiring IT skills was accurately captured.

Most of the interviews lasted between 20 and 90 minutes and were conducted either in the ODK office or in cafes nearby. Interviews were arranged a few days in advance, with a few exceptions when women were interviewed during or just after the workshops. All of the participants signed consent forms and gave permission for the interviews to be recorded. Most of the interviewees explicitly stated that they do not wish to be anonymised and that they would like their names to appear in the paper. Interviews were conducted in English and in order to facilitate freedom of interaction, no notes were taken during the interviews. Furthermore, throughout the duration of the fieldwork my daily presence in the office and at various events organised by GCK and ODK enabled me to establish warm, individual relationships with the staff. The interview recordings were transcribed and coded using a mix of descriptive and in vivo coding to organise the material conceptually. Subsequently, the codes were grouped into themes and categories on the basis of their frequency and structures of meaning and were then compared with existing theories.

The main limitations of the methods used in this research are inherent in many qualitative approaches: particularly the difficulty of applying the findings to wider populations due to context-specificity, and the implicit subjectivity in the interpretation of the findings (Yin 2008). Moreover, there are limitations in the robustness of the data collected: due to the majority of participants of the programmes pursuing their studies at the time of the courses, there was no employment impact data available. Furthermore, since the participants were interviewed within the context of the courses, their criticism of the programmes might have been tempered by their surroundings, thus leading to potential biases in the data. Another limitation is that the interviews were conducted in English, which is not the native language of the participants.

Findings and analysis

Disempowerment

The main sources of disempowerment as reported by the interviewees derive from the traditional divisions of gender roles, with women in many, particularly rural, families being discouraged from undertaking employment. 17 out of the 20 research participants perceived cultural constraints as the main reason for low female labour market activity, with ten participants explicitly having been told that ICT is not for women at some point in their lives. The perception that women should stay at home and raise children is still prevalent. Many women are actively encouraged to live off their husbands' earnings: 'they are glad they don't have to work. They believe it's a privilege' (Kosovare). 'Discrimination starts not when they go to work, the first point is families' – reported one of the interviewees, raising the issue of social norms being so deeply engrained that women believe that the optimal choice is not to work.

At university and in the workplace, direct discrimination occurs when women's skills are openly undermined, and their achievements are discredited: 'This is not a product you developed – you are just women, you can't do that', heard one of the interviewees from her professor. Dhurata, who owns a tech start-up at the age of 23, recalls being at an international conference and talking to people interested in her product, when she was asked: 'Where are the guys who did this?' The answer that she developed the app with her friend was followed by a series of technical questions aimed to verify whether she was telling the truth. Such scepticism and playing down of women's abilities prevail in the workplace. Kosovare, one of the most experienced women interviewed, claims that it is hard for people to believe that a woman can develop a successful programme or an app. This attitude results in women eschewing ICT and developer roles and gravitating towards more administrative tasks: 'They are faced with too many people who try to stop their way, they hear that they are not able to do that because they are female, and they get demotivated'. Other, more subtle forms of discrimination, entail giving women tasks perceived as easier, for instance front-end development, rather than more challenging back-end; or assuming that women are better with design because of their gender: 'They try to give you easier jobs because you're a girl. They always give us front-end or design, but not all girls are good at design.' This leads to frustration and fear of not reaching one's full potential: 'I want to do the work you do, I didn't come here to get a better treatment,' stated one of the participants. Moreover, the interviews revealed that women are at a disadvantage in the industry from the very beginning because when they are younger they are not encouraged to engage with ICT-related activities. Men, on the other hand, start programming from a very young age and are more experienced by the time they attend university. Given that the majority of interviewees reported that skills taught at universities are obsolete, such differences can significantly influence the careers of women.

The obstacles created by unfavourable socio-cultural norms contribute to lowering women's self-confidence and negative self-perception, which then negatively impacts their ability to persevere in ICT: 'Not many girls go into programming because it was installed in our heads that it's very hard and girls can't do it', says Kosovare. Although women's participation in the ICT industry is low partially due to such discouragement, the ones who chose it as a career emphasise that they are prepared to be challenged and are actively working to contradict similar stereotypes.

Thus, the research confirms the literature in discovering that the main source of disempowerment in the IT sector in Kosovo is socio-cultural norms (Curto & Simler 2017; Kelly et al.) strengthening the perception of ICT as a male domain, followed by the lack of female-only support networks (Diakonidze et al. 2016). However, this study provides additional insights on the subtle form of discrimination within high-skilled ICT jobs that have not been considered in the literature. Although McKay (2006), Hafkin and Taggart (2001) and Huyer (2006) are concerned about the feminisation of low-skilled jobs and advocate for women to become creators and developers of ICT products, they don't account for a form of discrimination explicit in this study, which is the allocation of easier tasks to women already working within high-skilled ICT professions.

Second, this research demonstrates that although insufficient university skills are a hindrance for all young Kosovars, women are particularly affected due to the fact that they are not encouraged to engage with ICT from a very young age in the manner that men are. Women often lack experience and the technological savvy that men gain early in their lives, which further magnifies disempowerment. Although the literature identifies low self-confidence and lack of self-esteem as significant sources of psychological disempowerment (Mishra 2016; Valarmathi & Hepsipa 2014), the interviewees are certain of their abilities. What constrains them is discrimination and exclusivity of male professional circles, which has been identified as a key obstacle by Franklin et al. (2005), Henwood (2000) and Cockburn (1983).

Technological empowerment

The ICT skills trainings organised by GCK and ODK leverage the obstacles described above as follows. First, they provide relevant skills demanded in the job market, thus creating a source of comparative advantage for women. This is of significance particularly due to dissatisfaction with the curriculum at universities in Kosovo. 15 interviewees reported that the skills taught at universities are obsolete, too theoretical, disconnected from their real-life projects and aspirations. 'Recent graduates don't have the skills that the market looks for – technology moves very fast, it's hard to get jobs straight away', says Dhurata, recalling the period of sadness and disappointment when having graduated with very good grades, she was struggling to find a job. She points to her GCK training as the factor that eventually enabled her employment in the sector. Moreover, by bridging this university-industry skill gap and training women in the latest technologies as well as soft skills such as teamwork, project management, communications, the courses provide them with the tools to become better programmers. Participating in the programmes prepares them to enter the job market with a potential advantage over their male counterparts who lack similar training. Furthermore, the mentors, women in their late 20s, 'digital natives' themselves (Prensky 2001), relate to the learning needs of the students and adjust their methods accordingly. These findings support the research by Nath (2006), Gurumurthy (2004) and Jorge (2002) that gaining new skills boosts participation in economic and social processes, as well as facilitates employment and increases women's capability to overcome potential obstacles.

In terms of the relevance of skills taught, the findings affirm the insufficiency of skills taught at universities (Diakonidze et al. 2016; Shala & Grajcevci 2018), with a majority of interviewees stating that what they were taught during their degrees was obsolete and insufficient to succeed in the job market. Moreover, the findings show that GCK and ODK staff ensure that the skills they teach are up to date through a careful process of curriculum design and industry research. Open data use is a significant part of the training, resulting in participants being fluent in practical use of open data and its tools at the end of the programmes. GCK and ODK teach technologies that, at the time of writing, cannot be learned anywhere else in Kosovo for free, which opens the courses up to more disadvantaged women.

Due to the quality and specificity of the skills taught, ODK and GCK often employ past participants of the courses, thus contributing to the labour market empowerment of women on a small scale. Five out of eight interviewees who now work for the organisations (founders are excluded from the calculation) got the job through participation in the trainings.

Social and psychological empowerment

By bringing women in ICT together, the programmes create a strong sense of community and introduce younger women to mentors and role models. This is a crucial empowering factor that has the potential to counterbalance oppressive social norms. The majority of participants stress that having a community of women in technology had a positive effect on their career development and confidence. Being in touch with other participants and organisers creates new friendships and provides practical assistance with ICT-related problems encountered during professional life after the programmes. As one of the organisers explains, in addition to teaching ICT skills, the aim of the courses is to 'motivate women to do anything they want to do, to challenge this stigma that men should do certain things and women should do others'. Past participants often call the organisers for technical advice. Some friendships that started during the programmes resulted in joint projects, such as a start-up developed by Argeta and Dhurata. Although the literature indicates that participation in such networks aids job searches (Stoloff et al. 1999; Dustmann et al. 2016; Zinnbauer 2007), only a few interviewees found employment through referrals. However, with GCK and ODK being recognisable and valued by the local employers, involvement with the trainings resulted in favourable treatment in the industry for the majority of interviewees.

Communities of women willing to support each other outside of the trainings were formed through bonding, as described in the literature (Storper 2005; Sanyal 2009; Pitt et al. 2006; Maclean 1999; Karim 2008). However, the findings add another dimension to the argument that cross-community bridging creates mutual trust (Gaved & Anderson 2006; Williams 2005;

Blanchard & Horan 2000). Code4Mitrovica, the programme that aimed to bring Serbian and Albanian communities to code together, resulted in a fruitful collaboration. Most of the Albanian and Serbian participants intended to stay in touch afterwards. However, they revealed that this consensus was based on a mutual willingness to avoid political conversations due to the potential for disagreement, thus demonstrating the shallowness of those new social ties.

Providing successful role models is another way in which the programmes empower young women in the ICT sector in Kosovo. Participants in their early twenties look up to their older colleagues, particularly Blerta, the founder of GCK, as a source of inspiration: 'You can put yourself in her shoes, you can imagine yourself doing this too' (Albana). Similarly, Kosovare, one of the mentors, had an older friend who supported her emotionally when she was starting to code. Now she provides similar assistance to the participants: 'It's a very good feeling when you do something for someone that someone did for you, helping you create your path'.

Some of the interviewees praise the effects the programmes had on their mental health and overall well-being and feel extremely grateful for having been given a chance to participate: '[Before I applied] it was a very low point, I was confused, sad. I felt very fortunate to be chosen, I really needed that experience. I learned so much' (Argeta). Similarly, Florina describes commencing the programme as a life-changing event: 'Things weren't going well. Getting into the training was the best thing that could happen to me – I hit rock bottom, I was kind of depressed; this was a new opportunity'. These findings show the significance of support networks for younger women on the verge of giving up due to the lack of promising job prospects. Here, guidance from older colleagues was particularly effective in shaping their future development.

Therefore, the research confirms Lennie (2002) and Mishra's (2016) suggestion that psychological empowerment contributes to a rise in self-confidence and self-esteem, and increases in motivation, enthusiasm, self-value, and well-being. The findings add to the literature by identifying the presence of relatable, young female role models as a key feature to evoke all these positive externalities.

Political and civic empowerment

As a young, post-socialist democracy, Kosovo struggles with corruption and low female participation in political life. Through data-driven projects focused on civic and political engagement, ODK and GCK create a platform for young women to engage with issues in their municipalities. ICT skills, open data, and civic engagement are intertwined in Kosovo. The majority of the projects created during the programmes use open data in the applications or websites they develop, and the topic of the trainings is always related either to improving the lives of residents or influencing the political climate. Therefore, such use of open data and ICT skills empowers women as citizens, raising awareness of their democratic privileges and encouraging their interest in male-dominated politics. Moreover, it teaches them to utilise new skills for greater social benefit. Increasing civic engagement is a significant aspect of the programme for the organisers: 'I want to ensure that women understand their role as active participants and know how to use technology to actually raise their voice', says Blerta. ODK and GCK promote the model of tech-enabled, active citizenship, where data are used as evidence to pressure governments and demand policy changes.

Working on open data-driven solutions results in a rise in female civic engagement. The interviewees reported an increase in their interest in local current affairs after the programmes: 'Before, I didn't care that much. I've started seeing things differently after the project because I worked with that. I think that reporting something that is not right in the city is important and the opportunity to have a meeting with municipality officials gives people the feeling they are being heard' (Artira). Women who were previously unaware of the civic aspects of technology are now taking the initiative to change things in their municipalities: 'I wasn't interested in open data that much; civic tech wasn't important to me at all. After I got to know it, I've started to learn new things about the process and it helps me develop my knowledge of different things, not only programming. Now, I look at how my municipality organises data on their website and I don't like it. It's not user-friendly, so people don't look at it. I want to voluntarily help them with their website' (Florina). Similarly, Arbenita reveals how working with open data changed her perception of its use: 'Before, if the municipality posted a document, I wouldn't even know about it. Now, I would go on their website to check it.'

What these findings show is that merely bringing women together is insufficient to stimulate civic or political engagement – in the case of GCK and ODK, involving them with open data and civic technology projects is what enables them to develop their interest in these spheres. As identified by Brandusescu and Nwakanma (2018), having such communities is essential: a lack of strong groups working on issues related to open data is one of the main obstacles in fully harnessing the potential of open data for the empowerment of women. This research shows that thus far, the programmes have made a significant contribution to creating a data-driven, civic community. Furthermore, through the projects' focus on a direct collaboration with local governments and on the design of e-governance platforms for their use, GCK and ODK help overcome the criticism of Noveck (2009) that such communities lack transformative powers by being disconnected from governmental practice. Here, women set the direction for the digital development of their governments through designing and building the tools they need. Hafkin et al. (2003) and Nath (2006) assume that providing previously unavailable information – open government data in this case – in addition to mobilising women around mutual concerns, can spark their interest in current affairs. However, this research demonstrates that for the majority of interviewees working in ICT, politics is not a primary interest and despite open data contributing to the shift in their perception of these issues, the increase occurs primarily in the civic, not in the political sphere.

One example of civic empowerment is as follows: for women participating in the programmes, gaining expertise in the use of open data and the opportunity to work on digital tools for their local governments gives them the confidence to speak up and liaise with their municipalities. Past participants of the Tech4Policy project describe the chance to express their opinions about issues regarding the municipality and having their advice taken seriously by the officials as the most fulfilling aspect of the programmes. Similarly, one of the project assistants observes: 'At the beginning, they lacked self-confidence to be engaged in this type of conversation with the municipality. It's because of the stereotype that municipality is hard to reach, and young citizens don't really deal with it. By the end of the project they were so interested, and knew much more about their municipalities, they grew as citizens'.

Women who previously did not work with municipal officials directly now perceive open data as facilitating their interactions with the public administration, for instance when obtaining documents or certificates from their municipalities. 'Open data is making our lives easier. Instead of going to the government's building and asking for the documents or data you need, you can just find it online' says Argeta. For her, the relationship between opening up the data and increasing transparency in governance is apparent: 'The government should be more open about their actions and open data is making it happen'. 'We should really be using it more often. Open data is free, you can make great things with it,' confirms another participant. Thus, the majority of interviewees recognise the advantages that the use of open data presents and its empowering potential. However, they are still uncertain how to use it in order to generate long-term social benefits.

General limitations

Despite their numerous advantages, ICT skills trainings in Kosovo have a number of general limitations that hinder their potential to fully promote the empowerment of women.

First, although the programmes provide the skills necessary to succeed in the workplace, they offer little practical employment assistance. As explained by one of the organisers: 'We don't help them find jobs, we teach them skills to be ready to apply'. Sometimes representatives of local tech companies are invited for the final presentations as an opportunity to hire some of the participants. Nevertheless, despite sharing occasional job opportunities, GCK and ODK fail to provide any further employment help. Considering cultural constraints and the relatively hostile climate within the industry, there is a risk that women trained by the programmes might fall into administrative or low-skilled ICT jobs instead of utilising the advanced skills they acquire through these programmes.

Although in theory the programmes are free and open to everyone, they tend to attract a certain profile of university-educated (usually in an ICT-related subject) urban women who are already empowered enough to be able to find such programmes and express their interest in participating. Due to the courses being advertised primarily on social media, women with no presence on those channels and no internet access are unable to find out about the opportunities. Therefore, ICT-enabled empowerment is only applicable to a certain group and does not challenge existing social divides. Through targeting women primarily from the ICT sector, the programmes are complementary to already existing skills, refining and preparing young female talent to enter the job market equipped as well as possible, rather than training women with no ICT background with an interest in entering the industry. It can be argued that this approach facilitates finding better jobs for people who would have jobs regardless, and a more sustainable solution in terms of overall female empowerment would be to target the most impoverished and vulnerable groups, such as women from rural communities or ethnic minorities.

Similarly, civic technology and open data are only familiar to a limited group of people, usually from the ICT industry. When asked about the awareness amongst the general public, the interviewees confirmed that the majority do not know what it is or how to use it – older generations in particular: 'People don't know they are capable of influencing those issues. They should ask more about what is happening to our money, why is the political situation in Kosovo like this' (Rita). This confirms Gurnstein's (2011) scepticism that tools like open data and civic technology can operate successfully only when there is awareness and skills sufficient to utilise them. Therefore, the key to the successful use of open data and civic technology for the increase in civic and political engagement lies in expanding their accessibility and educating citizens. Furthermore, as outlined in the previous section, even though the trainings increase civic engagement, their contribution to political empowerment is insufficient. Although the use of open data creates a desire to be more politically active, the participants report a lack of tools and guidance on how to achieve it. Perhaps introducing workshops focused specifically on enhancing female political participation through the use of open data and civic technology would be beneficial for further developments in this sphere.

Open data is a relatively new concept in Kosovo, thus in order for its meaningful and effective adoption, organisations like ODK and GCK need to overcome a number of obstacles identified by the interviewees, such as the lack of skills to effectively utilise the opportunities that open data presents. When

Georges started ODK, the skills he needed in order to support the open data movement were scarce. He explains that providing comprehensive training is lengthy and costly, and retaining employers is challenging, considering the higher salaries offered in the private sector that attract workers initially trained by ODK. Thus, there is a need to further popularise the broader use of open data for it to emerge as a norm, rather than a rare activity available primarily to the people from the ICT industry.

Low digital advancement of local governments is another obstacle hindering the potential of open data to stimulate participatory policy-making and crowdsourcing of policies. Thus, ODK and GCK are working to improve digital infrastructure in the public sector. Dafina, the deputy director of ODK, describes the process as follows: 'We combined technology, people and open data. We realised that the government doesn't have the infrastructure to do anything, so we trained the community, raised their awareness, demonstrated that they can build digital solutions using open data – this way people can have more access, more data'. Currently, the organisations train the community in the active use of open data and motivate them to utilise the existing data creatively. One of the projects that emerged through the open data-centred training was Ec Shlirë (Walk Freely),[3] a mobile app created during a series of workshops with 30 young Kosovar women. In Kosovo, 47% of people are subject to harassment but from 2013-14, only seven cases were reported.[4] The app allows users to anonymously report sexual harassment that they have been subjected to and provides them with data analysis tools in order to highlight trends and patterns of street harassment in Kosovo. Data produced by the app is openly available and disaggregated by the type of harassment, location, perpetrators, age group of the survivor. Available for Android since February 2016, the app has been downloaded more than 1,000 times and has collected nearly 400 reports (Open Data Kosovo 2018). For some of the women participating in that project in 2016, it was the beginning of their data and ICT careers: for instance, Kosovare received a job offer at ODK following her participation in the workshop. Similarly, a few other participants describe working on Walk Freely as their first comprehensive encounter with open data.

Furthermore, despite local governments being eager to begin the process of opening up their data, there are still substantial gaps in the data currently available. This presents the potential bias in data sharing – the government might allow harmless, yet less important data to be shared, while simultaneously preventing the access to other datasets that might be of higher significance for organisations like ODK. For instance, as reported by the organisation leads, obtaining water and air quality data proved extremely challenging. Similarly,

---

3   http://iwalkfreely.com/
4   https://www.canadainternational.gc.ca/croatia-croatie/eyes_abroad-coupdoeil/EA-Croatia-RSH.aspx?lang=eng

the Business Registration Agency was reluctant to open its data. As a response, ODK scraped the data themselves and built their own business registration search engine, Open Businesses, with data from over 170 000 businesses in Kosovo, making it possible to find businesses by the name or the owners, as opposed to by number only in the official registrar by the Kosovo Business Registration Agency. Although this shows the resilience and resourcefulness of the open data community in Kosovo, the reluctance in sharing key data is a serious obstacle that could further hinder the potential of open data to tackle corruption and inform activism, and in the long run, empower women in the socio-political sphere.

Discussion and conclusion

The research questions explored in this chapter were:

1. What factors disempower women in the ICT sector in Kosovo?
2. What are the strengths and limitations of ICT skills programmes and open data for the empowerment of women?

The investigation aimed to analyse the use of ICT to improve the situation of women while simultaneously creating high-skilled, more politically active human capital.

Broader context

This section situates ICT skills trainings in a broader context and confirms Wilson's (2015) and Prügl's (2017) claim that, in addition to the empowerment of women on the individual level, female-focused developmental initiatives create a number of wider socio-economic benefits.

As demonstrated in Figure 3 below, broader social benefits deriving from empowering individual women in the four spheres introduce a number of positive changes, such as promoting active citizenship and the use of open data, initiating community reconciliation, and creating high-tech female human capital that could influence social norms in the long run. Therefore, despite their limitations, ICT skills trainings serve as a holistic developmental strategy with an end-goal of creating a highly specialised and democratically engaged female population. In the case of Kosovo, where public sector ICT systems are obsolete, university skills insufficient, levels of corruption and female unemployment high, such programmes mobilise the untapped potential of the female workforce and narrow the university-industry gap. The rise in civic empowerment is enabled by the use of open data during the programmes. Thus, ICT skills and open data have the highest empowering potential when used in conjunction.

Figure 3. Benefits from empowering women

	How GCK and ODK enhance it	Benefits for individual women	Benefits for the society
Technological Empowerment	Teaching women advanced ICT skills; Retention of past participants – employment on the small scale.	Increased technological capacities; Specialist skills facilitate finding employment.	Bridging the university–industry skill gap in the workforce; Increased number of highly skilled women in IT has a potential to challenge preexisting social norms and stereotypes.
Social Empowerment	Creating a community of women in tech.	Participation in a strong support network; Female-only network could assist with finding employment.	Strong social capital provides basis for civic engagement in the long run; Technology can bridge communities.
Psychological Empowerment	Strengthening the ties between women in the community; Providing more experienced role models.	Guidance, help, inspiration.	Increased self-worth and overall well-being of women makes them more likely to participate in the labour force and could contribute to the rise in national productivity.
Political/Civic Empowerment	Using open data to design apps and platforms; Matching problems in the local governments that could be solved using digital solutions with skilled female programmers from the area.	Entering the civic and political spheres that were previously out of their scope.	Citizens are more active in shaping democratic processes; Improved digital infrastructure in the public sector; Increased awareness of open data could help fight corruption.

Moreover, enhanced female civic technology involvement has been utilised to modernise digital infrastructure in the public sector (Figure 4), particularly through projects like Tech4Policy or Techstitution.

The findings correspond with Cornwall's (2016) framework on the process of change (Figure 1). ICT skills programmes in Kosovo provide initial access to resources and opportunities, and an increase in the number of women in the ICT industry has a potential to transform informal cultural norms and exclusionary practices in the long run. The fact that GCK and ODK programmes require women to design ICT tools for their municipalities, advocate for open data and transparency as well as closely collaborate with officials triggers systemic changes in the formal spheres of governance and public policy, allowing women to take a more active role in these previously male-dominated areas.

Figure 4. Enhancing female civic technology involvement

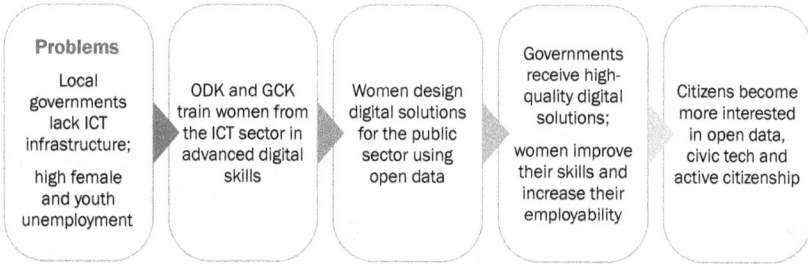

Problems	ODK and GCK train women from the ICT sector in advanced digital skills	Women design digital solutions for the public sector using open data	Governments receive high-quality digital solutions; women improve their skills and increase their employability	Citizens become more interested in open data, civic tech and active citizenship
Local governments lack ICT infrastructure; high female and youth unemployment				

Thus, this chapter argues that for ICT skills programmes to achieve sustained and multidimensional empowerment of women while simultaneously generating broad socio-economic benefits, such trainings should be implemented in the context of civic technology, the use of open data, and collaboration with local policymakers.

Summary

The popularisation of ICT in the development agenda has inspired the emergence of female-only ICT skills trainings aimed at empowering women and enabling their active participation in the knowledge economy. Through combining the existing definitions of empowerment, this research established a framework to assess to what extent ICT skills programmes contribute to the empowerment of women in four spheres: technological, social, psychological, political. The results indicated that although the programmes positively impact these aspects of women's lives, their potential to eliminate long-term sources of disempowerment is debatable due to the fact that they operate within patriarchal social norms.

These findings affirm the literature describing such norms as the main obstacle to empowering women, with discrimination, discouragement, and scepticism emerging as other hindrances. Overall, the research discovered that ICT programmes equip women with highly demanded skills, involve them within the support networks, positively influence their confidence and civic participation, using open data as a key tool to achieve the latter. Some limitations of the programmes were also identified: exclusivity, insufficient employment assistance, failure in strengthening women's political engagement.

Policy implications

Although the findings indicate that ICT skills trainings have the potential to partially empower women in the four domains (technological, social, psychological, political), it is crucial to acknowledge the limitations of the trainings as outlined throughout this chapter, and design policy solutions to overcome them. Thus, this section discusses the wider implications that should be of interest to policymakers in countries and regions with low rates of female and youth employment and civic and political participation.

First: Open data and ICT skills trainings should be deployed as complementary tools for upskilling the workforce and instilling the notion of active citizenship. Without relevant ICT skills, the presence of open data does not contribute to the empowerment of women. On the other hand, in order for the employment-focused ICT skills trainings to contribute to a greater social benefit, it is recommended for open data and civic technology projects to be their key focus. As proven by the case of Kosovo, inviting ICT practitioners to engage with open data increases their interest in municipal affairs and encourages their active participation in local decision-making.

Second: ICT skills training could provide a neutral environment for people from different – even historically antagonistic – communities to meet and collaborate. Such tech-enabled bridging could be reproduced in other locations in order to initiate the process of reconciliation and/or integration of ethnic minorities. Working towards a common goal – in this case designing a digital solution to improve the lives of inhabitants of Mitrovica – might not stimulate political discussions, yet it has the potential to unite younger generations on a professional level and establish projects they could work on collectively. This could lead to bridging and reconciliation between communities in the long run.

Third: In order for ICT skills programmes to produce long-term benefits for women, overcoming subtle discrimination in the workplace is essential. This could be achieved through establishing training-to-industry links for highly skilled graduates of the programmes to ensure that their career trajectories are not interrupted by placement in low-skilled, administrative occupations.

Limitations and future research

This study recognises that due to the methodological limitations, the findings shall not be treated as prescriptive. Qualitative studies are perceived as more subjective and interpretative (Yin 2008), and their implications are often context dependent. The data collected captured the ICT-enabled process of empowerment on the individual level. Thus, in order to overcome these limitations, further research is needed. Quantitative data on how many women found and retained employment in the ICT sector upon the completion of the programmes would create a comprehensive picture of technological and social empowerment. Further examination of the political empowerment achieved through the use of open data could provide additional insights on how to build capacity and stimulate political and civic engagement in the ICT sector.

Acknowledgements

First and foremost, I would like to express my gratitude to Blerta Thaçi and the rest of the Girls Coding Kosova and Open Data Kosovo teams. Thank you for sharing your experience, welcoming me to your country, introducing me to the Kosovar culture, and all your indispensable insights that contributed to this chapter. Further, I would like to thank Mor Rubinstein for her help and guidance, as well as Dr Neil Lee and Dr Susanne Frick for their advice and suggestions.

REFERENCES

Ananiadou K & Claro M (2009) 21st Century skills and competences for new millennium learners in OECD countries. *OECD Education Working Papers* No. 41. Paris: OECD Publishing

Ashcraft C, McLain B & Eger E (2016) *Women in tech: The facts (2016 Update). See what's changed and what hasn't.* NCWIT'S Workforce Alliance. https://www.ncwit.org/sites/default/files/resources/ncwit_women-in-it_2016-full-report_final-web06012016.pdf

Australian Bureau of Statistics (2002) Social Capital and Social Wellbeing. Discussion Paper. http://www.oecd.org/innovation/research/2380806.pdf

Bailur S, Masiero S & Tacchi J (2018) Gender, mobile, and development: The theory and practice of empowerment. *Information Technologies and International Development* 14: 96–104

Batliwala S (1993) *Empowerment of women in south Asia: Concepts and practices.* New Delhi: FAO-FFHC/AD

Batliwala S (1994) The meaning of women's empowerment: New concepts from action. In: G Sen, A Germain & LC Chen (eds) *Population Policies Reconsidered: Health, Empowerment and Rights.* Cambridge, MA: Harvard University Press. pp. 127–138

Bimber B (2000) The study of information technology and civic engagement. *Political Communication* 17(4): 329–333

Blanchard A & Horan T (2000) Virtual Communities and Social Capital. In: GD Garson (ed.) *Social Dimensions of Information Technology: Issues for the new millennium.* Hershey, PA: Idea Group. pp. 6-21

Boland L & Coleman E (2008) New Development: What lies beyond service delivery? Leadership behaviours for place shaping. *Local Government, Public Money and Management* 28(5): 313–318

Bowles S & Gintis H (2002) Social capital and community governance. *Economic Journal* 112: 419–436

Brandusescu A & Nwakanma N (2018) *Is open data working for women in Africa?* World Wide Web Foundation. https://webfoundation.org/docs/2018/07/WF_WomanDataAfrica_Report.pdf

Cockburn C (1983) *Brothers: Male Dominance and Technological Change.* London: Pluto

Cornwall A (2016) Women's empowerment: What works? *Journal of International Development* 28(3): 342–359

Davies T (2010) Open data, democracy and public sector reform: A look at open government data use from data.gov.uk. Unpublished MSc dissertation, Social Science of the Internet, University of Oxford. http://www.opendataimpacts.net/report/wp-content/uploads/2010/08/How-is-open-government-data-being-used-in-practice.pdf

Davies T (2012) Supporting open data use through active engagement. Position paper. W3C Using Open Data Workshop

27

Diakonidze A, Raja S & Gelvanovska N (2016) *Kosovo Digital Economy: Skills for Jobs.* Washington, DC: World Bank Group. https://mzhe-ks.net/repository/docs/Kosovo_ Digital_Economy-Digital_Skills_for_Jobs.pdf

Dustmann C et al. (2016) Referral-based job search networks. *Review of Economic Studies* 83(2): 514–546

Edwards R, Franklin J & Holland J (2003) Families & social capital: Exploring the issues. *Families & Social Capital ESRC Research Group Working Paper* No.1. London: South Bank University

European Commission (2016) *A New Skills Agenda for Europe.* Brussels. https://eur-lex. europa.eu/legal-content/EN/TXT/HTML/?uri=CELEX:52016DC0381&from=EN

Felten EW et al. (2009) Government data and the invisible hand. *Yale Journal of Law & Technology* 11(1). https://digitalcommons.law.yale.edu/cgi/viewcontent. cgi?referer=https://scholar.google.com/&httpsredir=1&article=1048&context=yjolt

Friedmann J (1992) *Empowerment: The Politics of Alternative Development.* Cambridge: Blackwell

Gaved M & Anderson B (2006) The impact of local ICT initiatives on social capital and quality of life. *Chimera Working Paper* No. 2006-6. Colchester: University of Essex

Granovetter M (1985) Economic action and social structure: The problem of embeddedness. *American Journal of Sociology* 91(3): 481–510

Gurstein M (2011) Open data: Empowering the empowered or effective data use for everyone? *First Monday* 16(2): 1–9

Gurumurthy A (2004) *Gender and ICTs.* Brighton: Institute of Development Studies

Hafkin N & Taggart N (2001) *Gender, information technology and developing countries: An analytic study.* Washington, DC: Office of Women in Development, Bureau for Global Programs, Field Support and Research, United States Agency for International Development

Hafkin N et al. (2003) *Engendering ICT: Ensuring gender equality in ICT for development.* World Bank Global Information and Communication Technologies Department. Washington, DC: World Bank Group

Heeks R (2008) ICT4D 2.0: The next phase of applying ICT for international development. *Computer* 41(6): 26–33

Henwood F (2000) From the woman question in technology to the technology question in feminism: Rethinking gender inequality in IT education. *The European Journal of Women's Studies* 7(2): 209–227

Hilbert M (2011) Digital gender divide or technologically empowered women in developing countries? A typical case of lies, damned lies, and statistics. *Women's Studies International Forum* 34(6): 479–489

Hussain F & Amin SN (2018) 'I don't care about their reactions': Agency and ICTs in women's empowerment in Afghanistan. *Gender & Development* 26(2): 249–265

Huyer S (2006) Understanding gender equality and women's empowerment. In: N Hafkin & S Huyer (eds) *Cinderella or Cyberella? Empowering Women in the Knowledge Society.* Bloomfield, Connecticut, USA: Kumarian Press

Jain S (2015) ICTs and women's empowerment: Some case studies from India. Department of Economics at LakshmiBai College, Delhi University

James L (2013, 3 October) Defining open data. Open Knowledge Foundation. Available at: https://blog.okfn.org/2013/10/03/defining-open-data/

Jorge SN (2002) The economics of ICT: Challenges and practical strategies of ICT use for women's economic empowerment. Prepared for the Expert Group Meeting on Information and communications technologies and their impact on and use as an instrument for the advancement and empowerment of women, Seoul, Korea, 11-14 November

Kabeer N (1999) Resources, agency, achievements: Reflections on the measurement of women's empowerment. *Development and Change* 30(3): 435–464

Kabeer N (2011) Contextualising the economic pathways of women's empowerment: Findings from a multi-country research programme. Pathways Policy Paper. Brighton: Pathways of Women's Empowerment RPC

Karim L (2008) Demystifying micro-credit: The Grameen Bank, NGOs, and neoliberalism in Bangladesh. *Cultural Dynamics* 20(1): 5–29

Kozma R (2005) National policies that connect ICT-based education reform to economic and social development. *Human Technology* 1(2): 117-156

Lee JC (2004) Access, self-image, and empowerment: Computer training for women entrepreneurs in Costa Rica. *Gender, Technology and Development* 8(2): 209–29

Lennie J (2002) Rural women's empowerment in a communication technology project: Some contradictory effects. *Rural Society* 12(3): 224–245. doi: 10.5172/rsj.12.3.224

Lin N (2001) *Social capital. A Theory of Social Structure and Action.* Cambridge: Cambridge University Press

Maclean K (2010) Capitalising on women's social capital: Gender and microfinance in Bolivia. In: S Chant (ed.) *The International Handbook of Gendered Poverty: Concepts, Research, Policy.* Cheltenham, UK: Edward Elgar. pp. 569–574

Masika R & Bailur S (2015) Negotiating women's agency through ICTs: A comparative study of Uganda and India. *Gender, Technology and Development* 19(1): 43–69. doi: 10.1177/0971852414561615

McKay SC (2006) Hard drives and glass ceilings: Gender stratification in high-tech production. *Gender and Society* 20(2): 207–235

McLoughlin C (2011) What ICT-related skills and capabilities should be considered central to the definition of digital literacy? In: T Bastiaens & M Ebner (eds) *Proceedings of ED-MEDIA 2011-World Conference on Educational Multimedia, Hypermedia & Telecommunications.* pp. 471-475. Lisbon, Portugal: Association for the Advancement of Computing in Education (AACE). https://www.learntechlib.org/primary/p/37908/

Milazzo A & Goldstein M (2017) Governance and women's economic and political participation: Power inequalities, formal constraints and norms. Background paper for the 2017 World Development Report on Governance and the Law. World Bank.

Ministry of Economic Development Kosovo (2013) Electronic communication sector policy – Digital agenda for Kosova 2013 ÷ 2020

Mishra G (2016) The psychological facets of women empowerment at workplace. *International Journal of Recent Trends in Engineering & Research* (IJRTER) 2(11): 224–228

Nath V (2001) Empowerment and governance through information and communication technologies: Women's perspective. *International Information and Library Review* 33(4): 317–339

Nath V (2006) Empowerment of women through ICT-enabled networks. In: N Hafkin & S Huyer (eds) *Cinderella or Cyberella? Empowering Women in the Knowledge Society.* Bloomfield, Connecticut, USA: Kumarian Press. pp. 191–206

Ng C & Mitter S (2005) *Gender and the Digital Economy: Perspectives from the Developing World.* New Delhi, Thousand Oaks & London: Sage Publications

Noveck B (2009) *Wiki Government: How Technology Can Make Government Better, Democracy Stronger, and Citizens More Powerful.* Washington: Brookings Institution Press

Noveck B (2012, 5 July) Open data – The democratic imperative. Crooked Timber. http://crookedtimber.org/2012/07/05/open- data-the-democratic-imperative/

Oladipo SE (2009) Psychological empowerment and development. *Edo Journal of Counselling* 2(1): 118–126

Oliver T (2017, 12 May) An in-depth look at the gender gap in the tech industry. *Technically Compatible*. https://www.technicallycompatible.com/an-in-depth-look-at-the-gender-gap-in-the-tech-industry

Peixoto TC, Sifry ML, Mellon AJ & Sjoberg FM (2017) *Civic tech in the global south: Assessing technology for the public good*. Washington, DC: World Bank and Personal Democracy Press

Pitt M, Khandker S & Cartwright J (2006) Empowering women with micro finance: Evidence from Bangladesh. *Economic Development and Cultural Change* 54(4): 791–831

Pollock R (2018) *The Open Revolution: Rewriting the Rules of the Information Age*. A/E/T Press

Prensky M (2001) Digital Natives, Digital Immigrants Part 1. *On the Horizon* 9(5). Lincoln: NCB University Press. pp. 1–15

Prügl E (2017) Neoliberalism with a feminist face: Crafting a new hegemony at the World Bank. *Feminist Economics* 23(1): 30–53

Putnam RD, Leonardi R & Nanetti Y (1993) *Making Democracy Work: Civic Traditions in Modern Italy*. Princeton, NJ: Princeton University Press

Ruijer E, Grimmelikhuijsen S & Meijer A (2017) Open data for democracy: Developing a theoretical framework for open data use. *Government Information Quarterly* 34: 45–52

Sanyal P (2009) From credit to collective action: The role of microfinance in promoting women's social capital and normative influence. *American Sociological Review* 74(4): 529–550

Shala A & Grajcevci A (2018) Digital competencies among student populations in Kosovo: The impact of inclusion, socioeconomic status, ethnicity and type of residence. *Education and Information Technologies* 23(3): 1203–1218

Stoloff JA, Glanville JL & Bienenstock EJ (1999) Women's participation in the labor force: The role of social networks. *Social Networks* 21(1): 91–108

Storper M (2005) Society, community, and economic development. *Studies in Comparative International Development* 39(4): 30–57

Ubaldi B (2013) Open government data: Towards empirical analysis of open government data initiatives. *OECD Working Papers on Public Governance* No. 22. Paris: OECD Publishing

Valarmathi G & Hepsipa J (2014) *A study on psychological empowerment of women in Urapakkam, Kancheepuram District. EPRA International Journal of Economic and Business Review* 2(4): 66–69

Vidovic H et al. (2017) *Western Balkans Labor Market Trends 2017*. Vienna: World Bank Group

Weare C (2002) The internet and democracy: The causal links between technology and politics. *International Journal of Public Administration* 25(5): 659–691

Williams K (2005) Social networks, social capital, and the use of information and communications technology in socially excluded communities: A study of community groups in Manchester, England. Dissertation, University of Michigan

Wilson K (2015) Towards a radical re-appropriation: Gender, development and neoliberal feminism. *Development and Change* 46(4): 803-832

World Bank (2017) *Kosovo – Systematic Country Diagnostic (English)*. Washington, DC: World Bank Group. http://documents.worldbank.org/curated/en/282091494340650708/Kosovo-Systematic-Country-Diagnostic

Xavier FP & Joseph MV (2013) Women empowerment: The psychological dimension. *Rajagiri Journal of Social Development* 5(2): 163–176

Yin R (2008) *Case Study Research: Design and Methods*. London: Sage

Zimmerman MA (2000) Empowerment theory: Psychological, organizational and community levels of analysis. In: J Rappaport & S Seidman (eds) *Handbook of Community Psychology*. New York: Kluwer Academic/Plenum

2.

Journalists and the intermediation of open data: A Nigerian perspective

Patrick Enaholo & Doyinsola Dina

Over the years the open data movement has reached a remarkable and ever-growing number of countries. Several groups and actors have embraced the campaign towards the liberation and democratisation of data in ways that would allow it to be interpreted and shared through networked platforms such as the internet (Davies 2012). Although the origins of the movement can been traced to specific political and social activities that began in the 1990s, most notably in the UK and the US, it was the initial attempts at formulating a definition of *openness* that catalysed a wider appreciation of the concept among more diverse segments of society (Heimstädt et al. 2014). Many of these definitions pointed beyond emphasising the need for a steady supply of data (mostly from governments) to their use, reuse and redistribution (by citizens) with little or no restrictions (Open Knowledge Foundation n.d.). According to the World Bank data toolkit, the two dimensions of data openness are legal and technical openness. To be legally open means data must be placed in the public domain or under liberal terms of use with minimal restrictions. On the other hand, data must be technically open, that is, it must be published in electronic formats that are machine readable and non-proprietary, so that anyone can access and use the data with common, freely available software tools. Data must also be publicly available and accessible on a public server, without password or firewall restrictions Thus, the availability and accessibility of data served as a pre-condition for its openness along with its machine-readability – a quality that enables data to be re-ordered, re-shaped and re-transmitted using computer-based algorithms.

However, the fact that data is open does not necessarily imply that it is useful. The usefulness of data requires that some underlying meaning can be drawn from it in order to generate new knowledge or enrich existing knowledge for the benefit of its end-user. While open data may be understood as content that can be freely used, modified and shared by anyone for any purpose (Open Knowledge Foundation n.d.), it is often generated and made accessible by a supplier and processed and packaged in the most appropriate format that an end-user can benefit from. Thus, within an open data ecosystem, 'end-users' are the final recipients in the spectrum of data transfer that begins with the data supplier. In the context of open government data (OGD), governments are the suppliers or providers of data and citizens are the end-users. Along this spectrum of data transfer, studies have pointed to the existence of stumbling blocks posed on the side of data providers as well as data users. Data provider-centric barriers include the reluctance to openness as a result of: a) the effort and cost required to convert closed into open data; b) the potentially high cost of facilitating the uptake of complex datasets; c) the poor quality of datasets; d) the absence of legal and policy frameworks; e) a lack of capacity to implement and sustain open data practices; and f) resistance by data custodians to opening up datasets. Data user-centric barriers include lack of access, low levels of data literacy, lack of human, social and financial capital to effectively use open data, and open up several datasets that can potentially create greater value when combined together (Janssen & Zuiderwijk 2014; Magalhães et al. 2013; Gurstein 2011). González-Zapata and Heeks (2015) summarise the main barriers of OGD into four categories: data absence, lack of data provision, lack of data quality and the digital divide. The last item, according to the authors, is a 'recurrent obstacle in developing countries' (2015: 4).

These challenges in the transfer of open data have warranted the need for a group of actors who perform crucial functions of sustaining the data supply chain through advocacy, interpretation and communication of open datasets for easy digestibility. These actors are referred to as open data intermediaries because of their role as linkages between open data suppliers and users (Van Schalkwyk et al. 2016). Open data intermediaries work with the data suppliers and/or the end-users to overcome the barriers associated with the provision or utilisation of open data (González-Zapata & Heeks 2015).

One of the early studies of the role of intermediaries in open data ecosystems was carried out by Magalhães et al. (2013) who identified three terms commonly adopted in academic literature and communities of practice: civic start-ups, open data services and infomediaries. With developments in the study and practice of open government data ecosystems and intermediaries, these overlapping categories have been overtaken by a wider set of groups with boundaries that are more clear-cut and distinct. For example, Johnson and Green (2017) use the term 'infomediaries' more broadly and synonymously with 'intermediaries' to encompass those that make open [government] data more accessible and useful for end-users. Among the categories of infomediaries in their classification are government agencies, the private sector, non-governmental organisations and community groups (including civil society organisations (CSOs)), academics and researchers and the media.

Scholars have also highlighted the complementary roles played by open data intermediaries. González-Zapata and Heeks (2015) classify them into five different types of agents on the basis of their interaction with data. Thus, intermediaries can act as demanders, producers, validators, developers and communicators of open government data. While the first three roles are directly related to the production of data, the latter two are linked to the manipulation of data to create information. They also note that, in practice, these roles are not mutually exclusive and intermediaries may perform multiple tasks at the same time. In line with this argument, Enaholo (2017) highlighted research on the role of CSOs in the OGD ecosystem: either as advocates for the provision of government data (demanders) or as grassroots advocates for citizen participation in the use of data (communicators) or both.

To perform any of these roles, open data intermediaries must possess the skills and competencies required to transform data into formats that can be beneficial to end-users. On the basis of their varying competencies within the open data ecosystem, Van Schalkwyk et al. (2016) apply Bourdieu's field theory nomenclature of 'capital'. The authors distinguish the financial, technical, social, cultural and symbolic capital as part of the repertoire that enable open data intermediaries to perform their roles within the ecosystem. The authors further argue that while one advocacy group may possess the symbolic or cultural capital required to apply pressure on government to release open data (as demanders), they may lack the technical or social capital required to convert the data into information that can be effectively communicated to end-users. Inversely, intermediaries with technical or social capital may lack the requisite symbolic or cultural capital to facilitate the supply of data from the government. Crucially, Van Schalkwyk et al. (2016: 15) note that intermediaries do not typically operate as single agents but rely on the capitals possessed by other intermediaries to 'unlock the full value of the transaction between the provider and the user in each of the fields in play'.

This complementarity of roles can be demonstrated using the list of infomediaries from Johnson and Greene (2017) highlighted above. Government agencies can act as demanders of data since their proximity to the originating source of data invests them with some symbolic capital. As highlighted by Van Schalkwyk et al. (2016), this role can also be played by civil society organisations with close ties to the government (perhaps through the agencies) as a result of their social capital. The media, academics and researchers may possess the required technical capital to interpret the data as information to be effectively communicated to end-users by other intermediaries with creative or social capital. The latter can include the media, non-governmental or civil society groups.

Our reason for specifying these roles along with their concomitant competencies (or capital) is to draw attention to the fact that those competencies often emerge from the skills and expertise usually associated with the professional groups to which open data intermediaries belong. When those skills are lacking, open data intermediaries may try to obtain them by improving those knowledge areas within their profession or acquiring new expertise from elsewhere. For example, Enaholo (2017) discusses how civil society workers in Nigeria are 'metamorphosing into a community of open data enthusiasts, perhaps in the hope that, through open data, the effectiveness of their role as advocates for good governance would be enhanced'. This transformation, according to him, 'is driven by the proliferation of CSOs with the skills and knowledge of web-based open data systems and tools' (Enaholo 2017: 95).

In this study, we have taken a different approach by examining the work of open data intermediation by media professionals, specifically, journalists. This approach is based on the need to recognise the distinctions that exist among the roles played by intermediaries based on their professional biases and competencies. To this end, we draw attention to the role of journalists and demonstrate how their work of open data intermediation involves competencies that appear to tilt towards end-users. Thus, in the terminology used by González-Zapata and Heeks (2015), journalists can be recognised primarily as 'communicators' of information (generated from data) but also, to some extent, as developers. While the former necessitates the possession of creative competencies, the latter focuses more on technical ones. From this perspective, we examine the intricacies of this intermediation – the processes and barriers – and how it contributes to or hinders their end-users' ability to assimilate open data.

In their study of the context of open data in Nigeria, Mejabi et al. (2014) examine the motivations, capacities and sustainability of intermediaries' role. According to the authors, CSOs analyse and re-present budget information while media professionals interpret the budget data to drive civic engagement and participation. In our research, we extend the findings of Mejabi et al. (2014) by: first, examining in some detail the processes by which journalists (as media professionals) carry out this role; second, unearthing those factors that limit and undermine their ability to do so; and, third, determining how these factors affect the broad spectrum of intermediation within the open data ecosystem. In doing this, we also contribute to the body of literature on the nature and diversity of roles and capacities of open data intermediaries in developing contexts.

Conceptual framework

Journalism and the 'quantitative' turn

The journalism profession is one that is characterised by the activity of reporting factual events in society to an audience. Journalists engage in gathering, assessing, creating and presenting news and information for public consumption. They disseminate the narratives generated from this process through a variety of media channels, most notably print and online newspapers, but also radio and television. As members of a profession, journalists differ from bloggers and those people who employ the press tools they now have in their possession to disseminate information (known as citizen journalists) by the fact that they hold themselves to a higher standard. Besides gathering and disseminating information, journalists try to assess the accuracy of the information, take care to avoid inadvertent error, identify sources (and their motives) whenever feasible, and ensure that promotional materials, and photos, video, audio, graphics, sound bites, and quotations do not misrepresent the information (Davis 2010).

Coddington (2015) notes that although journalism has historically been built around textual and visual presentation of reportage, the use of numbers was traditionally less prominent until the emergence of computer-assisted reporting (CAR) in the early 1970s – the period which marked, as Petre (2013) suggests, the advent of journalism's 'quantitative' turn. Computer-assisted reporting is considered the first approach to the systematic use of computers for data gathering and statistical analysis in journalism (Gray et al. 2012) and was a direct descendant of precision journalism, a term that was famously formulated by Philip Meyer. Since then, computer-assisted reporting has brought about an assortment of computer-based journalism practices including programmer-journalism, open-source journalism, data journalism, computational journalism, among others (Coddington 2015).

Among these variants, data (or data-driven) journalism has been identified as the most prominent successor of CAR. It was born out of the apparent glut in the availability of data in recent times (Borges-Rey 2016) and has emerged as a more rounded form of journalism than its predecessor. Some have argued that data journalism should be seen as the result of continuity in the profession's foray into the realm of computers rather than an outright change in approach. However, the development of CAR was in a context of information paucity that prevented journalists from finding answers to pressing questions and contributed to the need to devote time to gathering and analysing data using computer-based methods (Gray et al. 2012). Computer-assisted reporting was therefore a technique for augmenting reportage rather than a core part of journalism.

On the other hand, besides using data as a means to enhance stories, data journalism places data within the journalistic workflow. Data is at the heart of the data journalist's task of transmitting news and information to the public. It presents a broader understanding of news reporting: from a narrow conception of 'news events' coverage to the broader notion of 'situational reporting' in such a way that journalists move 'beyond the reporting of specific and isolated events to providing a context which gives them meaning' (Gray et al. 2012: n.p). Indeed, as Schrock and Shaffer (2017: 4) argue, 'data can be approached as a medium because it is inscribed with meanings, transmitted, decoded, and interpreted through specific practices.' Data journalism may therefore be described as a subset of processes involved in using data as a tool in the task of uncovering the hitherto obscure meanings behind events and phenomena that occur in society.

The practice of data journalism

Essentially, the work of the data journalist is to explore datasets to unearth new stories or to support existing ones. Since these stories are often concealed in large datasets, journalists require a specific set of specialised skills in addition to the domain-specific expertise to fully understand the data and its potential impact (Rogers et al. 2017). Data journalism is therefore perceived as a technique to extract relevant information from such datasets with the aid of statistical, visualisation and interactive methods for analysing, clustering and presenting data (Aitamurto et al. 2011). For proponents of the continuity school of thought, data journalism also inherits the expertise of data mining and collection from computer-assisted reporting.

With increased availability and better tools, data journalism is driven by the need to explain medium to large datasets. In developing stories and reports, the data journalist becomes 'the gatekeeper to reducing complexity and providing meaning to the data' (Rogers et al. 2017). Accordingly:

> [data journalism] represents a new role for journalists as a bridge and guide between those in power who have the data and the public who desperately want to understand the data and access it but need help [in doing so]. (Rogers et al. 2017)

For this, data journalists must augment their journalistic skills with those that would enable them to work effectively with data for the benefit of their audiences.

In general agreement with other schools of thought, we identify three categories of skills required by data journalists: 1) the treatment of data as a source to be gathered, verified and validated, 2) the application of statistics to interrogate it, and 3) visualisations to present it. In addition to these can be added the skill of reporting which is already inherent in journalism practice. For this study, we adopt these categories but we break them into five individual competencies for the data journalist: a) data collection (or mining); b) data validation (or cleaning); c) data analysis; d) data visualisation; e) data reporting. These competencies do not cover all the steps taken by journalists in carrying out their work. Rather, we focus specifically on those ones that are directly related to journalists' manipulation of data in drawing out news stories. For example, some scholars (such as Bradshaw 2014) have suggested data verification as an important skill or process in journalism that supports the ethical standards that are key characteristics of the profession. This comprises the protection of data sources, the effective handling of data leaks and privacy. However, since these processes do not necessarily involve working with data, we do not include them among our list of competencies. We discuss each of these in turn.

Data mining

Even with the abundance of data in contemporary society, data journalists often need to search for what is relevant and accessible to them. This requires the ability to explore a variety of options such as freedom of information (FOI) requests, internet searches, or crowdsourcing techniques. Each of these processes demands a sufficient level of mastery to yield the desired outcome. For example, obtaining data from the internet requires a knowledge of modern search techniques, the ability to use web-scraping tools or to write simple code that can pull data from the internet. In addition, a basic understanding of the prevailing laws will be needed to take advantage of FOI requests.

Data validation

This is also referred to as data cleaning. At their raw state, datasets typically require cleaning and validation in order to be useful for any form of analysis. When data journalists acquire data from various sources, they often need to combine them using software such as spreadsheets or databases. For this, they need to be familiar with any of the available tools that perform this function. Occasionally, they also require a basic knowledge of statistical procedures in order to prepare the data for more robust analysis. This step is important because it helps guarantee the accuracy of findings from the data.

Data analysis

Analysing datasets is central to the data journalism process (Aitamurto et al. 2011). According to Doig (2012), it is akin to interviewing a live source: 'you ask questions of the data and get it to reveal the answers.' Data analysis, also referred to as data interpretation (Knight 2015), is the differentiating factor between CAR and data journalism, with the latter laying more emphasis on a more inductive and exploratory approach that does not necessarily depend on how the data was collected (Coddington 2015). To carry out data analysis, journalists need a basic understanding of data structures as well as working knowledge of numerical and statistical principles.

Data visualisation

The practice of data journalism places emphasis on graphics and visualisation in the presentation of news and information (Knight 2015; Rogers 2011). The importance given to visualisations is essentially bound up with the data journalist's attentiveness to the audience (Coddington 2015). Visualisations are graphical representations of stories and reports that make it easier to consume news stories and narratives transmitted from data. Effective visualisations are the result of graphic design skills along with the ability to display quantitative information using visual perception and cognitive principles (Tufte 1983). Data journalists make use of a variety of tools and techniques to create visualisations from spreadsheets to graphics editing tools to programming languages.

Data storytelling

This is the final stage of the data journalism process in which journalists use storytelling techniques to transmit findings from the data to their audiences. The writing of news stories is often within the standard skillset of journalists. However, data stories (i.e. stories within the data journalism context) are typically amalgamations of stories in textual form with visualisations in graphics form. Oftentimes, these stories include the raw datasets published for readers to analyse and explore for themselves (Gray et al. 2012).

It is important to note that the order of the stages presented here does not always match what obtains in reality. Indeed, there has been a debate in the practice and research of data journalism about the preferred workflow for journalists (Rogers et al. 2017; Uskali & Kuutti 2015). Aitamurto et al. (2011) note the existence of two primary workflows. In the first, journalists start working with data only after the discovery of a story idea. This is similar to CAR whereby the data simply serve only to complement the story. In the second workflow, data serves as the starting point for the whole story. This matches the order of stages presented above. There is a general agreement in the literature that the former workflow is more common among journalists, although Uskali and Kuutti (2015) opine that the latter will be the future of data journalism.

Data journalism as open data intermediation

From the foregoing, it may be inferred that when journalists operate within the confines of an open data ecosystem, they are more likely to do so as data journalists. Rogers (2011) refers to this practice as 'open data journalism', the difference being that, in this instance, journalists work with data that is open and democratised and has the characteristics described earlier, namely to be available, accessible and machine-readable. Since the main interest of this study is open government data, we focus on data journalists who assume the role of open data intermediaries because the data with which they carry out the processes outlined above flow from the government as primary suppliers to citizens as end-users. Our concern then is to understand how journalists – as data journalists – fulfil this role within the ecosystem.

To this end, we formulate a conceptual framework that combines theories around the competencies of open data intermediaries (Van Schalkwyk et al. 2016) with the five stages of the data journalism process highlighted in the previous section. Thus, just as Van Schalkwyk et al. (2016) suggest that multiple intermediaries in an open data ecosystem possess different forms of capital (or competencies), we show how the work of data journalism involves two main competencies – technical and creative – which stem from the five stages of their expected workflow. This is illustrated in Figure 1.

Figure 1. The data journalism workflow within an open data supply chain

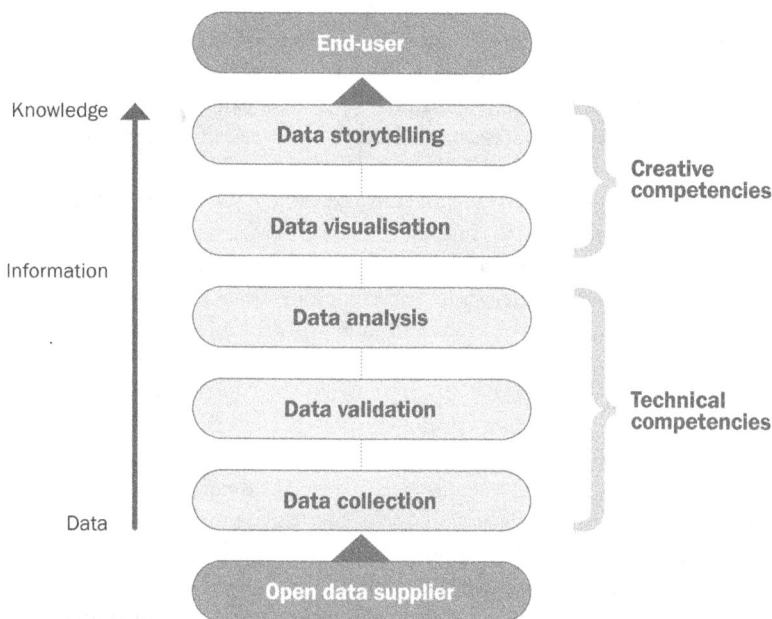

In the diagram, it can be seen that technical competencies comprise data collection, validation and analysis. These are the parts of the workflow that require more data-orientated skills from journalists partly because they involve the manipulation of raw data. By technical competencies, we refer specifically to the ability to handle data in a way that is initially removed from the information that they contain. These competencies are not exclusive to journalism. In fact, they can be said to belong primarily to disciplines with roots in statistical and quantitative analysis. However, the availability of tools and software have made data processing skills more accessible to journalists in the task of investigating and unearthing factual stories. With the technical skills required for data manipulation, journalists will be equipped to effectively interrogate datasets and use the answers derived to write reports.

On the other hand, the creative competencies are more attuned to the existing skillset of journalists who deal more with the art of relaying information in a clear and comprehensible way. According to Figure 1, these competencies include data visualisation and storytelling. As with technical competencies, our focus is not the content of the stories but their form. Creative competencies refer to the skills associated with constructing the narrative, including the language, the style and the writing techniques that contribute to the overall readability that makes it easier for audiences to understand what is being conveyed.

39

Methodology

In the sections above, we have drawn on a variety of sources to identify data journalism as the particular genre of journalism practice that pertains to the role of open data intermediary within the open data value chain. We have also outlined the processes (or stages) involved in the work of data journalists and shown how these can be embedded within the broader context of an open data ecosystem. Since there was a need to examine how this framework exists in practice, we engaged with data journalists in order to find answers to our research questions. Our aim was to identify those factors that enhance, limit and undermine journalists' ability to intermediate open data and how such factors affect the broad spectrum of intermediation within the open data ecosystem. The study is therefore data-centric as it focuses on the effectiveness of journalists' application of data as a tool to support news stories. For this reason, we do not examine the content of the news stories themselves since this would lead us away from the primary focus of the chapter.

Our study comprised three research methods: an online survey, content analysis and focus group discussions. We took the quantitative routes for the first two methods and qualitative for the last method. In total, 127 journalists from Nigerian mainstream media as well as freelancers who write for media houses were selected. Out of these, 94 journalists responded to the online survey using Google forms and 33 journalists joined the focus group discussions. In addition, 20 data stories that were shared by survey respondents were selected for content analysis by the authors.

Survey

The first research instrument used for the collection of responses was an online questionnaire prepared and distributed by email to the respondents. Purposeful sampling technique was adopted. The questionnaire was broadly targeted at journalists and sent specifically through data journalist networks. To further restrict the respondents to those who were more likely to be practising data journalism, we sent the survey as part of a recruitment process for a data journalism workshop. The survey was targeted at journalists based in a Nigerian newsroom or media outlet and this was verified by asking for the name of their media organisation along with links to their previous news stories if available.

The overall aim of the survey was to understand the application of data journalism among those who practice it. For this, we sought to find out what data journalists understood by the concept of 'data journalism' in order to determine whether it corresponded to the objectives of open data intermediation (i.e. to bridge the gap between open data suppliers and end-users). We also wanted to ascertain their level of awareness and practice of the processes involved in data journalism and to know the extent to which they applied those processes

40

when working directly with open government data (i.e. publicly available data disclosed by the government). In addition, the survey included questions about the common workflows that journalists adopted when working with data. This was to enable us to deduce how their preferred workflows affected their role as open data intermediaries.

Content analysis

After the online survey, we analysed the content of 20 data stories which were shared with us by survey respondents. The stories were selected from those that were provided by the respondents and were shortlisted using the following criteria: first, they contained some reference to numeric data; second, they were published in recognised media platforms; and third, they were provided by a journalist who demonstrated a clear understanding of data journalism and had practised it to a reasonable degree.

The choice of this method was informed by the need to verify the claims made by the respondents about their knowledge and practice of data journalism. Our assumption was that, if they did understand and apply the processes of data journalism in their work, it would be evident in their output (i.e. their data stories) – except if they encountered objective challenges and limitations in performing the various tasks involved in data journalism. Thus, we reckoned that the inability to work effectively as data journalists would negatively reflect on their role as open data intermediaries. If this was indeed the case, we wanted to see its manifestation. And to understand why this would be the case, we resorted to a focus group discussion.

Focus group discussion

A total of 33 journalists, most of whom have been working with data for about two to five years were selected for the focus group discussions (FGD). The method specifically helped to provide answers to our research questions dealing with those barriers and limitations that hindered journalists from using data effectively in their role as open data intermediaries. Our interpretation of effectiveness here refers to journalists' ability to undertake the processes required to convert raw data into information that can easily be digested by their audience (i.e. the end-users of the open data spectrum). For this, we specifically asked journalists about their possession of the skills needed at each stage of their work with data. This was in a bid to know whether they actually possessed the required competencies to carry out the various tasks involved and, if not, the challenges they encountered in acquiring or putting them into practice.

The FGD also served as a mechanism for triangulation, whereby the responses from the discussants were used to confirm or better understand answers they gave in the survey. This was important because, as we discuss below, some of the survey responses did not correspond with those we received during the FGD. For example, respondents' choice of data journalism workflows 'in theory' did not match what they carried out 'in practice'.

In our view, the methodological triangulation we adopted by using three different methods increased the credibility of the result and ensured a deeper understanding of the phenomenon we sought to investigate, more than would have been achieved with a single one.

Findings

From the survey responses we received on the concept of data journalism, we found that most of the respondents understood what it stood for and its relationship with the journalism profession. Many of the answers demonstrated a clear grasp of the notion of data journalism and, in their own words, described the relevance of their data stories to the audience. For example, one of them stated:

> My simple concept of data-driven journalism is journalism practice based on Data Analytics. That is, a system where a journalist engages the analysis and filtering of large datasets for the purpose of creating a news story. In doing this, he or she examines datasets in order to draw conclusions about the information the data contains, so as to present a credible news story to his/her readers or audience as the case may be. However, let me add that the importance of 'data-driven news story' cannot be overemphasised as it helps the general public or specific groups or individuals to understand patterns and make decisions based on the findings of the given journalists.

For the purpose of our examination of data journalism through the lens of open data intermediation, this response, like some others given in the survey, is useful because it specifically points to the end-user as the beneficiary of the journalist's attempt to 'draw conclusions about the information [contained in the data]' and presented in the form of 'a credible news story.' This clearly aligns with the role of open data intermediaries in transforming open data into information and packaging it into a digestible form (i.e. the news story) that end-users can benefit from.

The only drawback of this particular answer (and many others) was the lack of reference to the data provider. However, a few did include the data source in their definition and specifically referred to government data.

> My understanding of data driven journalism is [that it is] the new aspect of journalism where journalists [look] through figures around government or public funded projects to find the story the figures tell. This is about statistics, numbers help journalists to explain the narrative around a project, budget, sub-sectors, etc.

However, a different question in the survey gave the opportunity for the respondents to select their usual sources of data for writing stories (such as the World Bank, International Monetary Fund (IMF), Nigeria Stock Exchange (NSE), Central Bank of Nigeria and so on). Interestingly, the most common data provider was a government agency, the Nigerian Bureau of Statistics (NBS), which is known to publish datasets from different sectors of the country. Perhaps the omission of the data source in their definitions of data journalism implies that, since government data is the default provider of data, it can be taken for granted.

Beyond their understanding of data journalism as a concept, the survey sought insight into respondents' practice of it. Thus, it included a question about the processes of data journalism in which participants were most involved. Among the 5 skills listed in our framework, the most frequently selected was data storytelling. This was expected since, as mentioned previously, storytelling falls within the regular skillset of journalists as their primary occupation. Other skills included data analysis, data collection and data validation – in this respective order. The final skillset, data visualisation, was the least selected by the journalists. Even though we have classified visualisation under the creative competencies, it requires graphic design skills which can be technical and unfamiliar to journalists. As we discovered afterwards, journalists are often at a loss when trying to choose the most appropriate graphical representation (e.g. charts, infographics) to enhance the textual content of their news stories.

The lack of knowledge and use of data visualisation were apparent in the data stories analysed. A textual analysis of the stories revealed that, in most of the stories, numbers are discussed within the text, but not represented visually. This made it cumbersome to read them. For some of these, the numbers were represented as visualisations (often in the form of charts or infographics), thus repeating the same ideas conveyed in the text. Sometimes, these visualisations were not legible because they were too small thus rendering them useless since they added little to the reader's comprehension of the story. For some others, there was no clear correlation between the story itself and the visualisation, which had the overall impact of confusing, and perhaps misleading, the reader. An example is shown in Figure 2.

Only one out of all the stories analysed did not have any of the problems outlined above. Rather, it included charts that were legible and that complemented the story even though the information conveyed was sufficient for readers.

The fact that most of the data stories were difficult to read or comprehend goes beyond the poor quality of their visualisations. Gray et al. (2012: 191) argue that, 'unlike other visual media – such as still photography and video – data visualisation is deeply rooted in measurable facts'. This is because it is based less on aesthetics than on the data that it is designed to transmit. Since visualisations are information presented in graphical format using pre-attentive attributes that encode the underlying data (Few 2004), they depend on other aspects of the work with data such as analysis, validation and, to some degree, data collection and mining. Therefore, creating effective visualisations often requires some knowledge of the other three.

Figure 2. News stories with supporting data visualisations

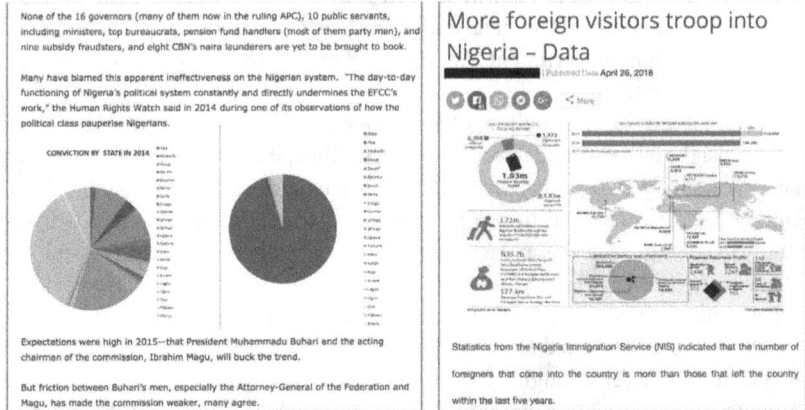

Source:https://nationaldailyng.com/efcc-fails-to-nail-single-politician-among-353-convictions-secured-in-10-years/

Source:https://www.dailytrust.com.ng/more-foreign-visitors-troop-into-nigeria-data.html

In the focus group discussions, we interrogated the reasons behind the poor quality of journalist's data stories and discovered that, contrary to claims about how they worked with data, the journalists lacked the expertise to actually manipulate the data using modern tools and software. We found that most of the knowledge they had was simply theoretical but often not practised. We also discovered that many of the journalists did not independently create visualisations to better their stories. Rather, they relied on charts, graphs or infographics created by others within their organisation or from external sources such as civic tech organisations. One of such organisations that featured frequently in our research was Code for Nigeria. Code for Nigeria is a civic tech group that uses data to give citizens hyper-local and hyper-personal information to make better informed decisions about everyday issues. They offer tools that give infomediaries like journalists and civic activists an easy 'plug and play' toolkit for finding and embedding interactive data visualisations into their storytelling. Indeed, some of the data stories analysed included visualisations created by Code for Nigeria.

Our research is not without limitations. For example, our sample size could have been larger in order to accommodate more journalists. We also restricted our focus groups to just one. Clearly, the study would have benefitted from a more ambitious recruitment of respondents even though we were limited by insufficient funding and logistical challenges – since journalists are not the easiest set of individuals to assemble in one place. In addition, all the data stories analysed were online. The inclusion of news reports published offline may have provided more nuance to the study. Finally, to increase the robustness of our findings, we could have included other actors in the open data value chain such as CSOs and end-users.

Discussion

Despite the findings and limitations, the fact that data journalists can and do operate as open data intermediaries cannot be denied. Therefore, how effectively do data journalists play this role? Based on our study, we hold that, in the Nigerian context, the contribution of data journalists is currently limited since they do not appear to possess the key competencies they need to communicate more effectively through their data stories. As shown above, our sample size of data journalists lacked the technical competences needed to act constructively as intermediaries despite their theoretical knowledge of their field.

While the creative competency of storytelling already forms part of their existing repertoire of skills (with varying degrees of proficiency), we found that journalists' lack of other skills required in their work with data – that is, the technical competencies of data collection, validation and analysis – obstructs their ability to work with open data. Although there exists a growing set of online tools that make it easier for journalists to use data for their stories, we argue that, based on our findings, these may be insufficient without the fundamental skills needed for data manipulation. While their lack of technical skills may be related to the absence of subjects like statistical analysis during their basic journalistic training, it can also be attributed to their lack of interest in moving beyond press releases and flashy headlines as well as structural problems in their media organisations or in society that lead to a lack of motivation or editorial mandate.

Nevertheless, based on the interactions with the journalists in our study, there was a general acknowledgement that their technical competencies were lacking or needed to be improved. Apart from the opportunities for skills acquisition which they frequently try to take advantage of (such as the workshop which was the bait for our research), journalists rely on the technical competencies of other actors in the wider ecosystem to complement their own creative competencies. These other actors can be individuals from within or outside their media organisation such as civic tech groups.

This kind of intervention supports research findings from authors such as Van Schalkwyk et al. (2016) and Enaholo (2017), among others, about the complementary roles played by multiple intermediaries in the open data ecosystem. In this particular scenario, the storytelling skills of data journalists (a key component of their creative competences) is combined with assets (in the form of data visualisations) provided by other actors like Code for Nigeria to compensate for the lack of technical competencies that journalists need in their work of intermediation. However, in spite of its benefits, our findings suggest that this collaboration is potentially fraught with shortcomings because its results were not optimal. This assessment is based on our observation of the output from the data journalists we researched. The quality of the written article did not often match that of the visualisations that were designed to complement them. For example, one of the stories, a well-written report by a journalist about the shortage of women in Nigeria's political space, included charts prepared by Code for Nigeria that were illegible and unclear, and that failed to effectively play the role of supporting the main article.

Although visualisations are not the only means of displaying data in news stories, they are most effective in concealing the difficulty usually associated with trying to comprehend the meaning behind a large amount of numbers in a text-based medium. For this reason, along with words, charts and visualisations complete the ensemble of a genuine data-driven story. We argue, therefore, that the poor use of data can be manifested by the low quality of data visualisations in terms of meaning, accuracy and complementarity with the main article. In our view, such cases of poor data visualisations are the outcome of a lack of or insufficient coherence among stages of the data journalism process. We opine that when adjoining stages of the process are disconnected, there is a negative impact on the data story and, hence, on the effective intermediation of the journalist in the open data ecosystem. Thus, when the technical competencies (i.e. the use of data) do not seamlessly support the creative ones (the language, style and flow of the article) in the data journalism process, there is a negative impact on the effectiveness of open data intermediation.

To deal with this, one solution is the upskilling of journalists in order to improve their technical competencies (Rogers et al. 2017). Rather than taking the technically orientated tasks away from journalists, a panacea to the challenge of ineffective intermediation among data journalists would be to upgrade their competencies in their area of deficiency. The need for training and developing data journalism skills has been one of the main challenges facing journalists in different parts of the world (see Appelgren & Nygren 2014 for Sweden and Hannaford 2015 for the UK). This research suggests that the same can be said for the Nigerian context. In our view, improving the technical competencies of journalists who work with data may have the effect of improving the quality of their output as data journalists to the benefit of end-users.

Another recommendation is to diversify the competencies of journalists within newsrooms in order to be able to form data reporting teams made up of differently skilled individuals. In contrast to the introduction of data experts as external actors within media organisations, this strategy involves a closer integration and complementarity of skills and would ensure that data-driven news stories are no longer produced by individuals but by teams of journalists whereby each one contributes on the basis of their core competencies. This approach to data journalism would require a re-configuration of newsrooms, one which would comprise a variety of journalists with different sets of competencies – technical, creative and so on. The existence of data teams in newsrooms is not novel. However, in the Nigerian context, based on our findings, it is a strategy that needs to be adopted more in order to improve the intermediary potential of data journalists.

Our findings suggest that journalists play a vital role as intermediaries through the interpretation of open data in formats that are more digestible for end-users. Their intervention lies at the intersection between the open data and the principles of data journalism – a burgeoning practice in the journalism space. The combination of both has shaped the process involved in the performance of intermediation by journalists. We describe this process as a five-stage process, namely, open data *collection, validation, analysis, visualisation* and *storytelling*. As we discussed, these steps are not novel and have been applied in data-driven reporting among journalists for quite a while (Coddington 2015; Gary et al. 2012; Rogers et al. 2017). However, we also show that they introduce a relatively distinct methodology into the work of open data interpretation which can (and indeed has) been adopted by other actors in the ecosystem. While the first three can be categorised under technical competencies or capital (Van Schalkwyk et al. 2016), the final two (especially the last – storytelling) may be classified as 'creative competencies' which belong firmly within the domain of journalism.

These competencies require specific skillsets which need to be possessed in order to advance the work of open data intermediation. The creative competencies refer to those skillsets that pertain to journalists' ability to write stories and, to some extent, visualise data. While this set of competencies was common to journalists in our research, we found that the other set of skills – the technical competencies of data collection, validation and analysis – was deficient. We also found that, to overcome this challenge, journalists rely on the technical competencies of other actors in the ecosystem and complement those with their own creative competencies. However, this solution proved imperfect because it resulted in inadequate and defective information for end-users. In other words, the quality of open data intermediation is lowered. This is because the five-stage process that is specific to data journalism forms a coherent unity in which one stage feeds easily into the next. When there is insufficient coherence between two adjoining steps, the process becomes less effective.

The solutions we propose to absence of technical competencies include, on the one hand, the skilling and upskilling of journalists in order to improve their technical competencies and, on the other hand, a reconfiguration of journalist newsrooms to accommodate a wider variety of competencies within data reporting teams. From the perspective of the overall open data ecosystem, the process of skilling can also be directed at other intermediaries such as CSOs through the improvement of their own creative competencies. However, even if the latter is achieved, the desired outcome could be adversely affected by the fact that CSOs would lack the institutional paraphernalia associated with the journalism profession.

REFERENCES

Aitamurto T, Sirkkunen E & Lehtonen P (2011) Trends in data journalism. *Hyperlocal*: 1-27. http://virtual.vtt.fi/virtual/.../D3.2.1.2.B_Hyperlocal_Trends_In%20Data_ Journalism.pdf

Appelgren E & Nygren G (2014) Data journalism in Sweden: Introducing new methods and genres of journalism into "old" organizations. *Digital Journalism* 2(3): 394–405

Baack S (2011) A new style of news reporting: Wikileaks and data-driven journalism. *Cyborg Subjects*: 1–10. https://www.ssoar.info/ssoar/handle/document/40025

Borges-Rey E (2016) Unravelling data journalism: A study of data journalism practice in British newsrooms. *Journalism Practice* 10(7): 833–843. https://doi.org/10.1080/175127 86.2016.1159921

Bradshaw P (2014) Data journalism. In: L Zion & D Craig (eds) *Ethics for digital journalists: Emerging best practices:* Abingdon-on-Thames: Routledge. pp. 202–219

Coddington M (2014) Clarifying journalism's quantitative turn. *Digital Journalism* 3(3): 331–348. https://doi.org/10.1080/21670811.2014.976400

Davies T (2012) How might open data contribute to good governance? *Commonwealth Governance Handbook* 2012/13: 148–150

Davis M (2010) Why journalism is a profession. In: C Meyers (ed.) *Journalism ethics: A philosophical approach.* USA: Oxford University Press. pp. 91–102

Dal Zotto C, Schnker Y & Lugmayr A (2015) Data journalism in news media firms – The role of information technology to master challenges and embrace opportunities of data-driven journalism projects. Paper presented at the Twenty-Third European Conference on Information Systems (ECIS), Münster, Germany

De Maeyer J, Libert M, Domingo D, Heinderyckx F & Le Cam F (2015) Waiting for data journalism: A qualitative assessment of the anecdotal take-up of data journalism in French-speaking Belgium. *Digital Journalism* 3(3): 432–446. https://doi.org/10.1080 /21670811.2014.976415

Doig S (2012, n.d) Basic steps in working with data. In: *Data Journalism Handbook* 1. https://datajournalism.com/read/handbook/one/understanding-data/basic-steps-in-working-with-data

Enaholo P (2017) Beyond mere advocacy: CSOs and the role of intermediaries in Nigeria's open data ecosystem. In: F van Schalkwyk, SG Verhulst, G Magalhães, J Pane & J Walker (eds) *The Social Dynamics of Open Data.* Cape Town: African Minds. pp. 89–108

Few S (2004) Eenie, meenie, minie, moe: Selecting the right graph for your message. *Intelligent Enterprise* 7: 14–35

Gill M, Corbett J & Sieber R (2017) Exploring open data perspectives from government providers in western Canada. *Journal of the Urban & Regional Information Systems Association* 28(1): 19–29

González-Zapata F & Heeks R (2015) Understanding multiple roles of intermediaries in open government data. Paper presented at the 13th International Conference on Social Implications of Computers in Developing Countries. Negombo, Sri Lanka, May 2015

Gray J, Bounegru L & Chambers L (2012) *The Data Journalism Handbook: How Journalists Can Use Data to Improve the News.* Sebastopol: O'Reilly Media

Gurstein MB (2011) Open data: Empowering the empowered or effective data use for everyone? *First Monday* 16(2). https://doi.org/10.5210/fm.v16i2.3316

Hannaford L (2015) Computational journalism in the UK newsroom. *Journalism Education* 4(1): 1–21

Heimstädt M, Saunderson F & Heath T (2014a) Conceptualizing open data ecosystems: A timeline analysis of open data development in the UK. Paper presented at the

Conference for E-Democracy and Open Government (CeDEM2014), Krems, Austria, 21 May

Heimstädt M, Saunderson F & Heath T (2014b) From toddler to teen: Growth of an open data ecosystem. *JeDEM-eJournal of eDemocracy and Open Government* 6(2): 123-135

Janssen M & Zuiderwijk A (2014) Infomediary business models for connecting open data providers and users. *Social Science Computer Review* 32(5): 694–711

Johnson PA & Greene S (2017) Who are government open data infomediaries? A preliminary scan and classification of open data users and products. *Journal of the Urban & Regional Information Systems Association* 28(1): 9–18

Knight M (2015) Data journalism in the UK: A preliminary analysis of form and content. *Journal of Media Practice* 16(1): 55–72

Magalhães G, Roseira C & Strover S (2013) Open government data intermediaries: A terminology framework. Paper presented at ICEGOV '13: Proceedings of the 7th International Conference on Theory and Practice of Electronic Governance, Seoul, Korea, 22 October. pp. 330–333

Mejabi OV, Azeez AA, Adedoyin A & Oloyede MO (2014) *Case study report on investigation of the use of the online national budget of Nigeria.* Open Data Research Network. http://www.opendataresearch.org/sites/default/files/publications/Investigation%20of%20Open%20Budget%20Data%20in%20Nigeriaprint.pdf

Metta S, Messina A & Italiana RR (2017) An end-to-end approach for delivering data-driven stories. Paper presented at the International Broadcasting Convention (IBC2017), Amsterdam, Netherlands, 14 September

Petre C (2013, 30 October) A quantitative turn in journalism? Tow Center for Digital Journalism. http://blog.chartbeat.com/2013/10/31/quantitative-turn-journalism/

Rogers S (2011, 28 July) Data journalism at the Guardian: What is it and how do we do it? *The Guardian.* https://www.theguardian.com/news/datablog/2011/jul/28/data-journalism

Rogers S, Schwabish J & Bowers D (2017) Data journalism in 2017: The current state and challenges facing the field today. Google News Lab

Schrock A & Shaffer G (2017) Data ideologies of an interested public: A study of grassroots open government data intermediaries. *Big Data & Society* 4(1): 1–10

Stalph F (2017) Classifying data journalism. *Journalism Practice* 12(10): 1332–1350

Stewart KN (1948) Review of *Journalism in the United States* by Robert W Jones. *The New England Quarterly* 21(2): 279–280

Tufte ER (1983) *The visual display of quantitative information.* Cheshire: Graphics Press

Uskali TI & Kuutti H (2015) Models and streams of data journalism. *The Journal of Media Innovations* 2(1): 77–88

Van Schalkwyk F, Chattapadhyay S, Cañares M & Andrason A (2016) Open data intermediaries in developing countries. *Journal of Community Informatics* 12(2): 9–25

3.

Using open data for public services

Miranda Marcus, Ed Parkes, Therese Karger-Lerchl,
Jack Hardinges & Roza Vasileva

Many public services in the UK are expected to deliver efficiency savings along with improved outcomes for citizens. At the same time, public service delivery is increasingly interconnected with many organisations from the public, private and charitable sectors. To ensure effective and efficient delivery, we need to understand these links and interdependencies. The changing nature of data presents new possibilities – data is moving from being scarce and difficult to process to being abundant and easy to use. There are increasing opportunities to harness its value for economic and social benefits. Open data is data that anyone can access, use or share (ODI 2017). It drives innovation by and for governments, individuals, businesses, start-ups and communities. We can see the potential open data has for transparency, economic growth and productivity in international open data for accountability initiatives (OGP n.d.), open data start-ups (ODI n.d.) and businesses improving processes with open data (Shadbolt 2015).

Open data is already being used to provide public services. For example, the release of travel and mapping information by Transport for London reduced the need for the public sector to build services for end-users and helped build a new market of digital mapping and travel services to meet people's needs. The government created space in the market for the private sector to innovate while monitoring the market to ensure that it was fair, equitable and so that the public sector could meet user needs where the market had not. This resulted in reducing public sector costs, improving services, creating jobs and increasing economic growth. Despite this success, there are still many cases where the government is building new services that could be delivered by the market. Open Data Institute (ODI) has been researching existing examples of open data helping deliver policies and public services with an aim to understand how they were built; what were the lessons learnt, as well as the economic and social benefits produced. ODI has engaged with public sector service design teams to test the outputs and produce training materials for public sector staff to encourage the use of this service delivery model and provide funding for organisations inspired by the training to experiment with new delivery models.

This chapter explores the use of open data in public service delivery and its potential for collaboration, joint problem-solving and open innovation. It highlights where open data has been released by public sector institutions and what effects it had on delivering public services. This chapter has drawn on the existing research from Nesta, including their Wise Councils report (2016), which explores case studies of councils using data in innovative ways. Additionally, they have developed tools for others to redesign services, such as the DIY Toolkit,[1] or more recently, their Data Analytics in the Public Sector (Copeland, Dragicevic et al. n.d.) draft guidance.

Additionally, the Organisation for Economic Co-operation and Development (OECD n.d.) released a comparative study looking at how open data can drive innovation in public service delivery. The study is focused on the potential benefits of open data, but it signals a need for tools to support people working in public service design and delivery to use open data more effectively.

The research questions considered in this chapter are:

- Where has open data helped deliver policies and public services?
- What are the commonalities in how they were built; the lessons that were learnt; and what were the economic and social benefits produced?
- Can the design of services be improved by creating and influencing tools that address the barriers to open and shared data public service delivery?

The research surfaces three high-level patterns of open data use in public services:

- Pattern 1 uses open data to increase access to services for citizens or organisations.
- Pattern 2 uses open data to plan public service delivery and make service delivery chains more efficient; direct beneficiaries are commissioners, managers and frontline public service workers.
- Pattern 3 uses open data to inform policy-making; direct beneficiaries are elected representatives, policymakers and citizens who want to influence policy.

In this chapter, we identify examples of each pattern and draw insights from their similarities. We go on to develop practical recommendations for a range of actors to support greater use of open data to deliver public services.

---

1 http://diytoolkit.org/

Methodology

Defining public services

Developing a working definition of public services helped us identify appropriate examples to examine in more detail. The definition used in this chapter is: *a public service helps groups of people to fulfil a need, the fulfilment of which is viewed to be in the public interest, and which public sector organisations recognise the need to provide.*

We developed and settled on this definition through workshops with public service delivery managers and the research team for several reasons. We wanted to recognise a definition of public service which was as broad as possible, so we could capture the wider effects of open data, for instance, where data collected and published in one part of the system was used by another part of the system at another place and time.

Our current definition focuses on the needs that individuals or groups of people have, as suggested by Cassie Robinson (2017), as we did not want to align our definition too closely with the legal responsibilities that public sector organisations have. We wanted to recognise that – particularly in relation to services that rely on external organisations for their delivery and are not necessarily totally funded by the government – there would not always be a distinct legal requirement for provision to a particular group.

Previous work on open data in public services

The focus of this chapter and the conceptual framework is primarily the UK public sector. During our research, we identified blog posts, reports and previous analyses on the use of open data in public services. There is a wide literature on the use of data in central and local government (Nesta 2016) and the impacts of open data on economic growth and innovation, but a relatively limited range focused specifically on open data and public services.

This chapter builds on Davies' work on open data and public sector reform specifically addressing his finding that open data can 'reduce friction in co-production of services between different levels of the state' to highlight and analyse its underlying models (2010: 5). The work of Susha et al. (2017) on data collaboratives points to a similar set of patterns to those identified in this chapter as outcomes of the use of data collaboratives. These are characterised as 'prediction and alerts (i.e. using data insights as early warning signals), needs-based planning (i.e. using data to learn about people's needs for aid planning), capacity building (i.e. using data to identify areas to improve government response), and monitoring (i.e. using data to track compliance with policies)' (Susha et al. 2017: 2695). This chapter develops these outcomes within a UK local public sector context, without focusing on the mechanism of data access.

Pia Waugh (2016), who has worked for both the Australian and New Zealand governments, has outlined some effects of open data on public services in a number of blog posts and presentations, including this summary:

Efficiency: proactively publishing data that is commonly asked for in an automated way frees up resources;

Innovation: once data is published, so long as it is published well and kept up to date, other people and organisations will use the data to create new information, analysis and services. This innovation can be adopted by the agency, but it also takes the pressure off the agency to deliver all things to all people, by enabling others to scratch their own itch; and

Improved services: by publishing data in a programmatically accessible way, agencies found cheaper and more modular service delivery was possible through reusable data sources. Open data is often the first step for agencies on the path to a more modular and API-driven way of doing things (which the private sector embraced a decade ago). I believe if we could get government data, content and services API enabled by default, we would see dramatically cheaper and better services across all governments, with the opportunity for a public ecosystem of cross-jurisdictional service and information delivery to emerge. (Waugh 2016)

In the UK, the ODI's work with the Environment Agency (ODI 2015a) has captured the benefits of releasing open data since the Environment Agency made their organisation-wide commitment in 2010. The impacts the Environment Agency identified at that point were:

- Helping the Environment Agency to achieve its core objective;
- Saving time and resources;
- Building external relationships and getting the user-voice heard;
- Improving data quality and public perception;
- Better understanding and managing risks around data use and publication;
- Bringing diverse teams together;
- Working more easily and efficiently with external partners; and
- Harnessing the power of the tech industry to make useful applications.

Another sector-focused study by GovLab (2014) examined the opportunity for open data use in the UK's National Health Service (NHS) in 2014. Along with exploring potential impacts of open data on the system, it set out a conceptual framework which could be used to help measure the impact of open data in the NHS. The report developed a logic model for open data in the NHS, described as: 'the use of certain kinds of inputs and data, by certain kinds of users, for certain kinds of activities, will achieve certain outputs and outcomes that indicate impact. Specific methodologies will be used to collect and measure indicators, helping to assess impact' (GovLab 2014). The report also made recommendations for taking the agenda forward in the NHS, including to 'develop an open health data ecology map, possibly using crowdsourcing, with a dictionary of all open health datasets used along with the variety of uses and users' (ibid.).

In summary, previous research has begun to explore impact areas experienced by the public sector in releasing open data. This chapter expands on this previous work by identifying particular examples and their shared characteristics to enable a more detailed knowledge of how the release of open data will make public service delivery better.

Approach

The project employed a combination of traditional qualitative techniques such as interviews and desk research, alongside the technique of visualising public services as ecosystems. Due to the complexity and non-linearity of the public service delivery process, identifying impacts and beneficiaries is not straightforward. Impacts might be indirect, hard to capture, or might take time to materialise. Visualising a public service as an ecosystem allows to identify and be explicit about the impacts and who is affected and helps understand different types of value exchanged within networks of data publishers and users. Therefore, these impacts become clear to the public sector and data holders, encouraging the further release and use of open data.

Visualisation brings sense to complex systems and issues across business and government. Government departments have created visualisations of ecosystems to help convey the overarching strategy and vision for an organisation (DWP Digital 2015). Maps of business processes and data collection help understand opportunities in government data collection (Rose 2010). Visualising and drawing are core approaches in service design and the UK Cabinet Office Policy Lab and academics such as Lucy Kimbell have pushed their use forward in a government context. Kimbell (2015) has provided a number of useful approaches to thinking about service design in the public sector and her methodologies have inspired the way in which we have conducted research for this project. There are a number of academic disciplines that visualise ecosystems and networks to bring insight to complex systems. In particular, a soft-systems methodology can be helpful in capturing people's involvement in a system – it is an approach to business-process modelling which is useful for general problem-solving and managing change. The system was developed to deal with 'soft problems' – those where there are divergent views and where organisations and people are key parts – with lots of similarities with public sector systems. In addition, value network analysis uses visualisation to understand complex economic systems.

In our research, we experimented with a technique from the soft-systems methodology called rich picturing. When exploring an issue, we ran workshops with public sector groups working with open data, to visualise the relationships between organisations, technologies and datasets.

Participant selection

The examples of open data-enabled public services came from three sources. First, there were those which were already known to members of the ODI through its global work with the public and private sector. A second source was a broad internet search and screening research organisations that had similarly been focusing on the open data agenda. This included organisations such as Open Knowledge, GovLab, the Knight Foundation and Sunlight Foundation, amongst others. In addition, we looked at open data portals with listed examples of use. Finally, we undertook a search of reports and other documents published on the topic of data use within the UK government, with a specific focus on open data.

This secondary research was supported by interviews with experts from the community. The interviews highlighted examples and helped to develop and expand our approach and explanatory framework for the rest of the project. The details of the examples were mostly captured in workshops with individuals who had a close working knowledge of the area. We spent the majority of the workshop time working with the participants to draw a rich picture of the example identified. In the process of developing the illustrations and drafting the descriptions of the examples, we reconfirmed the details with relevant individuals and, where appropriate, omitted commercially sensitive or otherwise confidential information at their request.

Findings

We propose that open data can be used in the public sector to help deliver services in three ways. We call these 'patterns' and have used them to group the examples in this work.

Figure 1 shows that the patterns occur at different points of the public service delivery, moving from direct impacts for the user (on the left) towards indirect impacts through better delivery and policy-making (on the right).

High-level patterns of open data use:

- Pattern 1 uses open data to increase access to services for citizens or organisations;
- Pattern 2 uses open data to plan public service delivery and make service delivery chains more efficient; direct beneficiaries are commissioners, managers and frontline public service workers;
- Pattern 3 uses open data to inform policy-making; direct beneficiaries are elected representatives, policymakers and citizens who want to influence policy.

Figure 1. Open data for public services taxonomy

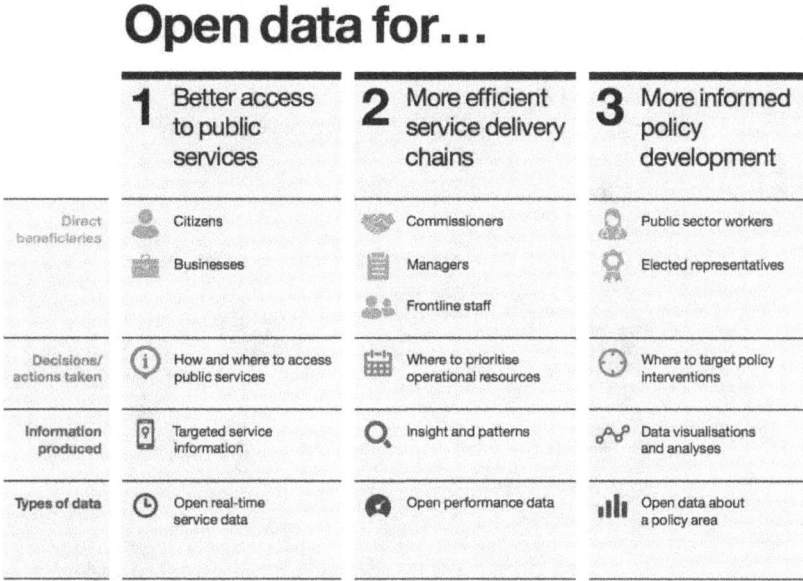

Open data for...

	1 Better access to public services	**2** More efficient service delivery chains	**3** More informed policy development
Direct beneficiaries	Citizens / Businesses	Commissioners / Managers / Frontline staff	Public sector workers / Elected representatives
Decisions/ actions taken	How and where to access public services	Where to prioritise operational resources	Where to target policy interventions
Information produced	Targeted service information	Insight and patterns	Data visualisations and analyses
Types of data	Open real-time service data	Open performance data	Open data about a policy area

Figure 1 shows how these patterns vary in terms of direct beneficiaries and differ in the types of open data currently released and used. For instance, in Pattern 1, we find more examples of rapidly changing data being used – such as that released by Transport for London (TfL) – than in the patterns targeted at civil servants (Pattern 3).

The impact is not limited to direct beneficiaries. It can extend to many, if not all, of those who deliver and receive public services. This report focuses on the benefits generated from open data. We recognise that open data can also create risks or feedback loops that have negative impacts and that potential outcomes should be carefully assessed on a case-by-case basis. The following sections look at the dynamics of the patterns in more detail. In presenting the ecosystems, we also explain how each was developed. This offers insight into how these arrangements came to be and highlights important aspects of how open data supports particular services that are not explicit when analysing the ecosystem drawings in a fixed state.

Pattern 1: Open data for increasing access to public services

The first is perhaps the most well-rehearsed pattern, since it clearly sets out an open data 'start-up' story where organisations delivering public services release data and start-ups use it to develop products or services. These examples tend to be of customer-focused, personalised public services.

Example 1a. Transport for London release of transport data

Transport for London (TfL) is a local government body responsible for the transport system in Greater London. TfL has the responsibility for London's network of principal road routes, for various rail networks including the London Underground, London Overground, Docklands Light Railway and TfL Rail, for London's trams, buses and taxis, for cycling provision, and for river services.

Figure 2. Transport for London data ecosystem

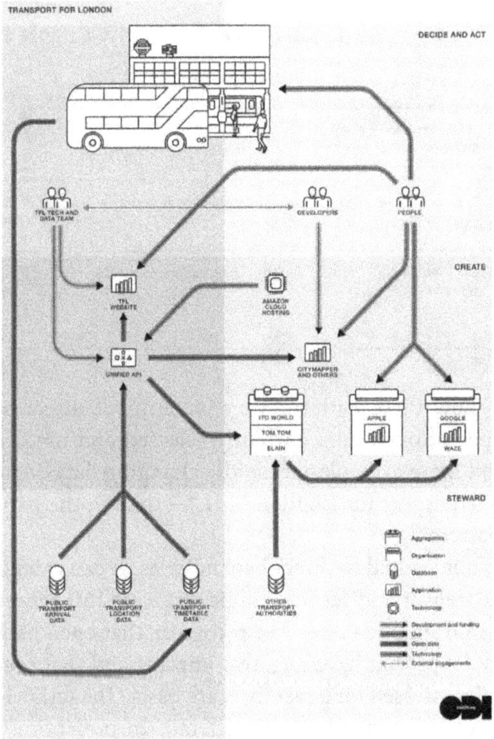

TfL is one of the UK's leading open data publishers. With over 31 million journeys made in London every day (TfL n.d.), it has long recognised the need to make travel information readily available to passengers. Publishing open data is a central part of TfL's customer information strategy of providing real-time data such as service locations, routes and delays to passengers far beyond their own online and offline channels.

Towards the end of the 2000s, TfL found that developers were scraping information about its services from its website. In an attempt to enable others to more easily display this information on their own websites and desktops, in 2007 TfL launched embeddable widgets – including maps of its network and live travel news. While TfL still makes a set of widgets available, the launch represented the beginning of a process in which the organisation would publish increasing amounts of data for others to access, use and share.

Between 2007 and 2011, TfL introduced an area for developers on its website and openly published real-time transit data via a range of feeds and downloads.[2] This helped to satisfy a growing demand for its data among developers, who used it to create user-facing journey planners and other applications. The anticipated influx of visitors to London during the 2012 Olympic Games was a stimulus for the introduction of live bus arrivals data, which led to a number of successful bus-only transport applications. Shortly after this came the launch of a new unified Application Programming Interface (API) for TfL's website.

The development of TfL's unified API in 2014 and the decision to open it up to external users was an important step in the organisation's open data journey.[3] Historically, the data it had published on different transport modes was made available in a variety of formats and structures, which made it difficult for developers to stitch together and develop multimodal applications (such as those that enable users to plan a journey using both buses and the London Underground). The unified API presented the data in common formats (XML and JSON) and, for the first time, consistent structures.

TfL's open data now covers timetables, routes and lines, embarkation points and facilities, transit status, disruptions and works, and fares.[4] According to recent research by Deloitte (2017), in total there are over 80 TfL data feeds (75% of which are available via the unified API) and over 13,000 registered developers.[5] Data users range from multinational technology companies to individual developers. The research has shown that TfL's approach to open data is improving journeys, saving people's time, supporting innovation and creating jobs.

According to Deloitte (2017), TfL open data is now used in over 600 apps (including journey planners, mapping tools, booking and scheduling tools, and analytics engines). 42% of Londoners use an app powered by TfL data and passengers benefit from between GBP 70 million and GBP 90 million per year in time saved from using open data-powered applications to plan journeys more accurately. Up to GBP 20 million additional revenue is generated from increased journeys per year, driven by access to travel information, and GBP 1 million is saved per year by enabling external development of new customer-facing

---

2   TfL also publishes open data related to transparency and accountability, such as details of its expenditure; this case study focuses on the organisation's transit data.

3   TfL, Unified API: https://tfl.gov.uk/info-for/open-data-users/unified-api?intcmp=29422#on-this-page-0

4   TfL Website (n.d.), 'Our open data', https://tfl.gov.uk/info-for/open-data-users/our-open-data?intcmp=3671

5   Note that TfL still supports the use of some data feeds and bulk downloads that preceded its unified API. In some cases, these contain additional data not yet available via the unified API.

apps, rather than producing campaigns, systems and apps in-house. TfL save GBP 2 million annually by moving away from SMS passenger alerts and TfL open data currently supports 730 jobs, including those in new companies made viable through its availability.

Insights

Clearly describing the benefits of open data in relation to TfL's wider organisational objectives has established a strong, ongoing case for its publication. Providing accurate and timely information to passengers is central to TfL's ability to deliver its physical services – open data enables developers and other organisations to develop customer-facing tools that do this on a scale that TfL could not do alone. Efforts to establish the value of the benefits of TfL's open data approach were documented in the Shakespeare Review (Gov.uk 2013) and the Deloitte (2017) research commissioned by TfL that has estimated the total value of open data to the organisation and customers to be of GBP 130 million per year.

Engaging with developers and other users has helped to build a vibrant ecosystem around TfL's open data. Having begun with creating an area dedicated to developers on its website in 2007, TfL continues to engage with and provide support to users of its open data through hackathons, blog posts on its website, and other channels. In 2017, it launched the TfL Tech Forum and ran a consultation focused on understanding how it could improve its open data publication. Publishing open data has supported conversations and partnerships between TfL and other organisations who hold valuable transit data. By making data available to others, TfL has benefited from engagement with companies such as Waze, Apple and CityMapper – some of which provide TfL with access to the data they collect. This includes data related to transport modes and areas for which it does not itself collect data (e.g. crowdsourced traffic incident data), giving TfL access to a rich source of data enabling them to better manage traffic in London.

A number of intermediary users combine TfL open data with data from other sources to create their own products and services. Organisations like ITO World, TransportAPI, Tom Tom and Elgin gather data from different sources to provide aggregated data feeds and additional services to developers and other organisations. This type of use demonstrates the need for transport data standards that allow interoperability – part of the value currently added by these intermediary users is to harmonise the data published by different transport operators and other organisations to make it easier to use and to develop new products and services.[6] As well as using open data published by TfL to increase access to the organisation's physical transport network, some

---

6   The ODI is currently working with a wide range of stakeholders to support the development of open data standards (http://standards.theodi.org). This includes producing a case study focused on the General Transit Feed Specification (GTFS), which is a widely adopted standard for public transportation data. https://docs.google.com/document/d/1m3jJR741VcE6ouyPITWgh 6HmASQG2jhfIMCddPAva8U/edit#heading=h.fzurpmmkvtj4

users are developing their own transport services. For example, CityMapper helps users make decisions about how to get from A to B in London using different modes of transport, which are managed and run by providers including TfL, Uber and Arriva. In 2017, CityMapper trialled the SmartBus, which uses data it collects on how users of its service move around the city to create the routes and timetables for the new SmartBus service.

Example 1b. Leeds release of data in relation to bin collection

The Leeds Bins app is a mobile application that tells people who live in Leeds when their green, brown and black bins are due to be collected, and adds reminders to their calendars. Open data on bin collection routes and times is used to inform people living in Leeds when their bins are collected. The application reminds the users to put their bins out the night before and includes links to what to put in which bin and where to take items that cannot be put into any of the bins. The reminders make the rubbish collection more convenient for citizens, and publishing open data instead of sending out letters saves the council approximately GBP 100 000 per year based on estimations by imactivate.[7]

Leeds has long seen open data as a means of supporting local economic growth while dealing with substantial reductions in local government spending power. In 2014, Leeds Council decided to invest significantly in making open data work for the city. Leeds has a lot of digital talent and the city leadership saw the potential to showcase the city's strengths to potential investors. In setting out the open data initiative, the council asked city departments what challenges they faced to consider how open data could help and assigned funding to encourage the release and use of open data. The Urban Sustainable Development Lab[8] set up to generate and pilot ideas was one of the programmes, with additional funding from the UK's national 'Release of Data' fund (Data.gov.uk 2014). Leeds' open data platform Data Mill North[9] hosted an event at ODI Leeds[10], an ODI Node, with developers working with council departments, including office staff, frontline workers, and elected members, to consider how open data could be used for better service delivery. People working in waste management shared the problems they faced in their work, one being that people did not know when their bins were collected and, as a result, were unhappy with the service. Furthermore, the council had to mail out bin collection timetables either annually or bi-annually, an expense that they increasingly could not afford. Open data on bin collection and other issues was made available for the ODI Leeds event and developers built prototypes to address the identified problems. Imactivate, a small software company in Leeds and a partner organisation of

---

7   https://www.imactivate.com/
8   http://sustainabledevelopmentlab.com
9   https://datamillnorth.org/
10 http://leeds.theodi.org/

ODI Leeds, developed a Leeds Bins prototype website. Four ideas in total were presented to the waste management department and the winning Leeds Bins team received the funding to develop what later became a mobile application.

Bartec Auto ID[11] has software that manages bin routes in Leeds and sends and receives live updates. Data Mill North worked with Bartec Auto ID to release household bin collection data openly, and the council included the open release of the bin collection schedule in its contract with the company (Data Mill North n.d.). The Ministry of Housing, Communities and Local Government (MHCLG) had previously developed a standard for publishing bin collection routing data with Bartec Auto ID, and other stakeholders, in their Local Waste Service Standards Project. Imactivate developed the Leeds Bins mobile application[12] using the open bin route data and the uptake of the application has been fast and widespread (Data Mill North n.d.). The council uses open data on app usage to identify areas of low uptake, and has launched targeted initiatives to increase awareness of the application and promote other ways of informing people when their bins are collected, to reduce the risk of people without smartphones not receiving information about collection times. These targeted initiatives are more cost-effective than regular mail-outs and the open data on uptake allowed Leeds City Council to choose which channels to prioritise. Data can inform these decisions yet, ultimately, the council needs to make them democratically.

Insights

Starting with the problem – in this case, people not knowing when to put out which bin – and working closely with the local authority on solving it with open data, has proven to be instrumental in delivering a better public service. The ongoing cost to Leeds City Council for this solution is about GBP 1 500 per year, significantly lower than the costs of mail-out.[13] Imactivate and Bartech Auto ID are now selling the app to other councils, powered by open data where possible and by direct data-sharing between imactivate and Bartech Auto ID where open data is not preferred by the local government, or where opening the data would result in unacceptable costs (e.g. PAF licensing).

Pattern 2: Open data for more efficient service delivery chains and planning

In this second pattern, we identify examples of open data release supporting service planning and delivery. The direct beneficiaries are likely to be those in public sector organisations, commissioning services and those in external organisations involved in service delivery.

---

11 https://www.bartecautoid.com/
12 https://datamillnorth.org/products/leeds-bins/
13 Numbers are based on estimations by imactivate.

These examples highlight the role open data can play in better coordination between the public and private or third sector organisations involved and make the delivery of public services more effective and efficient.

Example 2a. Local authority publication of spend data

Spend Network[14] pulls together spend, contract and tender data published by local authorities and other organisations in the UK. The organisation then provides insight and consultancy services to potential service providers to public sector organisations and provides similar services back to the government. Open data published by government organisations is analysed and repurposed by Spend Network to provide analysis, which can help make delivery chains and relationships between suppliers and buyers more efficient. The same (closed) data is used to save public money and improve the quality of services delivered to citizens.

Figure 3. Spend Network data ecosystem

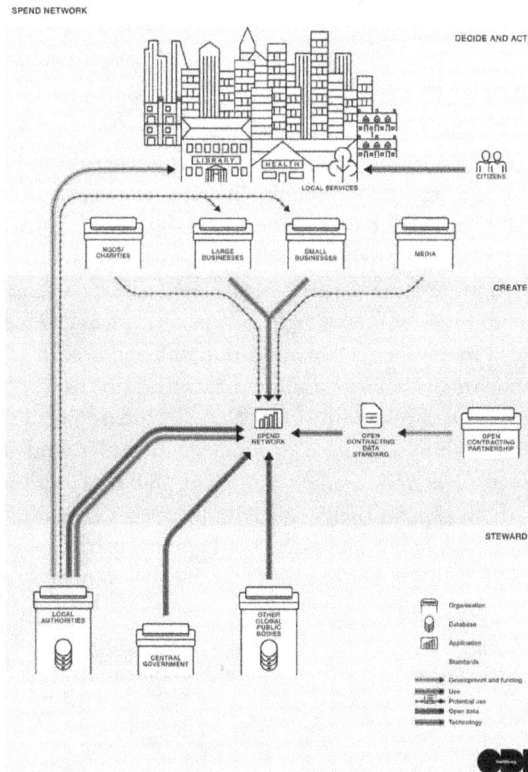

---

14 https://www.spendnetwork.com/

Inspired by the Windsor and Maidenhead Council[15] which had started to release details of their spending in 2008, the UK government, as a part of the transparency drive in the previous coalition government, set out expectations (Gov.uk 2015) of the spend and transparency data it needed the local authorities in the UK to publish (Rogers 2010). For instance, all UK Local Authorities are obliged to publish all spending transactions over GBP 500 and all Government Procurement Card spending and contracts valued over GBP 5 000. The policy motivation for these commitments was to ensure that taxpayers could gain insight into how public authorities were spending money. Supported by guidance produced by the Local Government Association (2015), some local governments began to publish this data in open formats and made data on let contracts available in addition. The councils were required to publish the data on their websites and Spend Network began to pick up the data and aggregate it across many different UK local authorities. The Network was launched in November 2013 using open data to create the first comprehensive and publicly available repository for government transaction data, a market worth in excess of GBP 130 billion per annum. Since then, it has published over 100 million transactions worth in excess of GBP 3 trillion. The company has grown out of Ticon, a small consultancy firm focused on government procurement and payments, founded by Ian Makgill.

Spend Network provides services such as procurement intelligence to both small and large businesses who are providers of services to local government. It is also involved in a data reseller market to large consulting firms who then offer services back to the public sector. Spend Network data is standardised and linked and can be used to compare between bodies, regions or sectors. The ability to compare can help with spotting patterns and anomalies, which can then be addressed to improve the delivery process. The data can also be used to compare between suppliers; for example, comparing prices can inform spending decisions and lead to more effective allocation of resources. In addition, Spend Network re-publishes the open government data made available according to the Open Contracting Data Standard (OCDS),[16] a standard created by the Open Contracting Partnership[17] and used in many countries around the world. The team behind Spend Network are now using the standard and their experience gained in the UK to expand their tender-finding services worldwide using the name OpenOpps[18] and their site is currently being used by Financial Times Stock Exchange (FTSE 100) companies and the government.

Insights
Despite the guidance from central government on the publication of data by local

---

15 https://www3.rbwm.gov.uk/
16 http://standard.open-contracting.org/latest/en/#
17 https://www.open-contracting.org/
18 https://www.openopps.com/

authorities, Spend Network devotes a lot of time and effort identifying, cleaning and analysing data. In this ecosystem they work as an aggregator – they curate the data, push for its publication when it is not available, query quality issues and suggest improvements. They also lobby for data with copyrights to be published openly. However, despite a large amount of activity from Spend Network and public sector organisations, we are yet to identify specific examples of where the data has been used to improve procurement in a public sector organisation.

Example 2b. NHS publication of open data

The NHS spends over GBP 110 billion a year delivering health services in England and has a complex arrangement for providing these services at a local level, with many organisations that support and oversee the design and provision of services. They often collect data which is used to analyse performance and improve quality and access to health services for citizens. The ODI undertook a workshop with members of NHS England and NHS Digital to begin mapping out some of the key open data use cases in the NHS. Nevertheless, the NHS is a vast network of organisations and understanding the full position of use of open data would take considerable further research and expert knowledge. The majority of data use within the NHS is individual-level data used for the delivery of direct care or, in pseudonymised or anonymised form, for research and planning. The data is made available to NHS organisations through the NHS Digital Secondary Uses Service[19] and to outside organisations through other means with appropriate controls in place. Some healthcare trusts have data-sharing arrangements with private, voluntary or academic sector organisations that can provide analytical insight and services to organisations at various levels within the NHS. Performance improvement organisations in the NHS such as clinical audits and the Commissioning Support Units (CSUs)[20] also use data and analysis in their work with healthcare trusts and Clinical Commissioning Groups.

We identified the key examples of open data use, including a number of crucial open datasets published (ODI 2015b). A relatively well-known example of the potential of open data in the NHS is the open prescribing data. NHS Digital publishes practice-level prescribing data[21] every month – this entails a list of all medicines, dressings and appliances prescribed by all practices in England, including GP practices. In 2012, Mastodon C and the Open Data Institute used this data to demonstrate the type of analysis that open data could provide. They issued a report highlighting the savings the NHS could make if they shifted from branded drugs to generic ones using the open prescribing dataset. Subsequently, an organisation called Open Prescribing[22] has been using the data

---

19 http://content.digital.nhs.uk/sus
20 https://www.england.nhs.uk/commissioning/comm-supp/csu/
21 https://digital.nhs.uk/practice-level-prescribing-summary
22 https://openprescribing.net/

to provide the ongoing analysis of the health service. There are further examples of open data publication within the NHS being associated with clinical impacts. For instance, when MRSA (meticillin-resistant Staphylococcus aureus, a type of bacteria resistant to several widely used antibiotics) instances were published as open data, there was an 85% reduction in the number of cases. However, it is difficult to disaggregate the impact that publication had from the influence of the media and quality and safety improvement work.

Insights

Although data is used routinely throughout the NHS, our initial research discovered that its potential is not fully realised. The few examples of open data publication are yet to demonstrate sufficient impacts on services. The NHS open data agenda could perhaps learn from the tactics and approaches used in ecosystems that we have explored elsewhere, where connections have been made between the potential uses of open data and its publication, and feedback loops established. We recognise that this is a more complicated undertaking given the complexity of the NHS systems and data.

Pattern 3: Open data for policy development

Finally, examples below centre around open data informing policy development and strategic direction. Here, the direct beneficiaries are involved in policy and strategy for the development of public services. Openness improves access to data across different branches of government and beyond, which can then be used to inform policy decisions.

Example 3a. The Department for Work and Pensions development of Churchill

Figure 4. Churchill

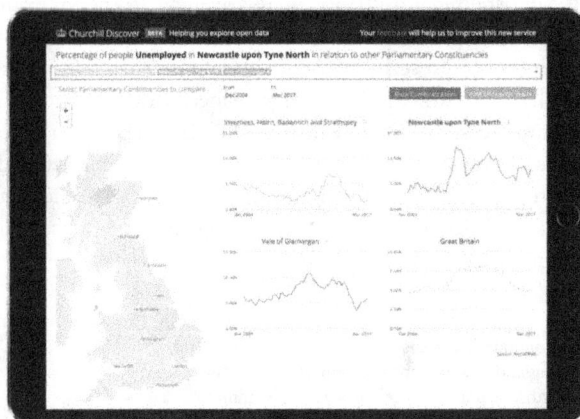

Churchill is a working title for a digital data service being developed by the Department for Work and Pensions (DWP) that is run entirely on open data APIs and data published in CSVs from UK government sources. The service is aimed at enabling DWP to understand their customer and claimant base better, through visualising trends within the UK over time such as pensions, labour markets, disability and health. It is a combination of a data visualisation tool built on D3 javascript libraries and a Mongo DB backend. Churchill helps policy and delivery officials in DWP to develop more evidence-driven policy and services with data visualisations pulled from open data across government.

Figure 5. Churchill data ecosystem

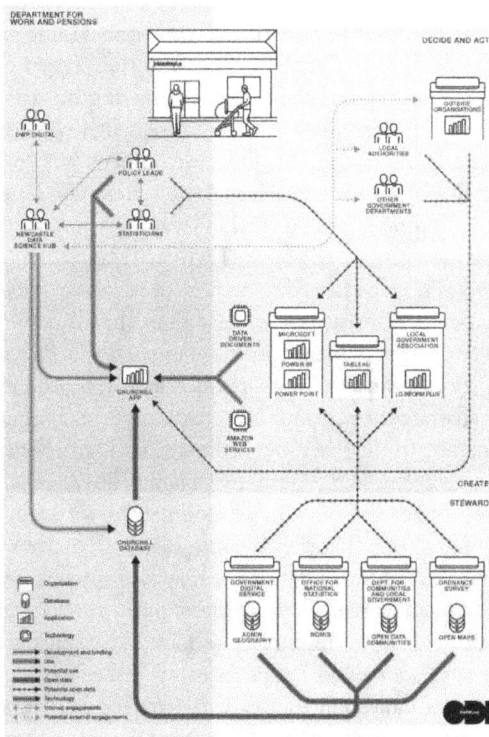

The DWP has a history of making its statistical data available to the public through publications and more recently through the online portal Stats-Xplore.[23] There was a renewed expectation that the department would become more digital and data driven (DWP 2017b), which resulted in organisational restructuring to bring data and digital teams together. The new Director General, Mayank

---

23 https://stat-xplore.dwp.gov.uk/webapi/jsf/login.xhtml

Prakash, made a commitment to 'driving hard for visualised analytics to be the norm, not the exception' (Prakash 2017).

One of the data science teams based in Newcastle began exploring the user needs of policy colleagues in DWP who use data to inform policy and design services. They undertook extensive user research with policy colleagues at Grades 6/7, developed a persona and elaborated the user needs for a new data-driven product. This work captured the level of data literacy as well as the current workflows and packages being used.

They also considered other products and services that could be used by policy professionals and analysts in DWP, such as LGInform[24] developed by the Local Government Association,[25] which pulls together data from DWP as well as data published by local authorities. The approach to the product was inspired by the world of data journalism and visualisations, including the ONS visuals site[26] which creates a number of statistical visualisations based on user research. DWP developed a prototype of the software drawing on data available from open data APIs across government, that is subsequently copied into a database and updated regularly. A key part of the development was to ensure that the platform satisfied DWP security protocols – the system was set up from scratch. The front-end visualisation is run on the D3 javascript library[27] and the product has been developed working in the open (DWP 2017a).

There were several reasons for using open data to power the service. First, in some instances, open data is easier to access than shared data.[28] Second, some of the interviewees perceived open data as more trusted as it had gone through a process of quality assurance prior to publication. The product was launched internally to demonstrate the possibilities to colleagues. Subsequently, Churchill has attracted interest from other departments and governments (Australia, Canada) and has been profiled externally (Darabi 2017). The main potential impact of the use of open data in the development of Churchill is likely to be the reduced amount of time that policy colleagues and analysts in DWP spend accessing, cleaning and analysing data. This may also support more iterative and agile policy development as issues can be investigated more quickly. Potentially, the tool could help release the resource of an analyst. Quicker data access and analysis can free up resources to better focus on policy and strategy decision-making. Besides, better data quality and visualisation enable to spot dependencies and relationships that can support more informed policy and strategy decisions, for instance, a more effective job seeker allowance policy. This is in line with the government's aspirations set out in the recent Government Transformation

---

24 http://lginform.local.gov.uk/
25 https://www.local.gov.uk/
26 https://visual.ons.gov.uk/
27 https://d3js.org/
28 ODI's Data Spectrum demonstrates the spectrum of data access between open and closed data, much of which is 'shared' data. https://theodi.org/about-the-odi/the-data-spectrum/

Strategy (HMG 2017) to make better use of data by making it available for internal uses through APIs.

Sharing the tool across the government would reduce the costs of each department procuring individual solutions. In a longer run, it could support greater standardisation of geographical data across the government as civil servants would be able to identify datasets that need standardisation through using the tool. It would also highlight anomalies or inconsistencies, which could help improve the quality of government open data and create a greater connection between data collection and use.

Insights

The release of open data has brought innovation within the government. In the case of Churchill, digital transformation tools and approaches supported the development of a data product. One of the key drivers was incorporating the system within a broader digital transformation agenda within DWP. However, further analysis is required to understand the potential improvements in the policy-making process following the release of Churchill, and how elected representatives and citizens could influence policy through this tool.

Discussion

This section will elaborate on the similarities identified across the examples discussed in previous sections.

Organisational collaboration

Each ecosystem analysed in this chapter is made up of organisations that all play a discrete role in the delivery of public services, as opposed to one organisation being responsible for the end-to-end process. Services in each ecosystem can only be delivered if a number of organisations and individuals have the resources and appropriate knowledge to do their jobs correctly. When working across organisations, the need for easy access and the use of data becomes even more important. Here, open data can be a key ingredient in ensuring greater coordination, more efficiency and a better end-user experience.

Technology infrastructure, digital skills and literacy

Another cross-cutting aspect of the ecosystems is the importance of technology. There is a minimum level of digital infrastructure, skills, literacy, and capability needed for organisations to make the most of open data in their public service delivery model. This relates not only to the technology within the public sector but also to the wider ecosystem of the private sector and civil society.

Foundational data infrastructure

Data infrastructure consists of data assets, the organisations that operate and maintain them, and the technologies and processes required to use the data (ODI n.d.). Strong data infrastructure creates new opportunities such as creating new products and services, developing new insights, and supporting diverse data ecosystems.

Often, data infrastructure includes foundational datasets that support a broad range of uses, particularly if made available as open data. There are several key types of datasets that support the different ecosystems we have identified. Given the nature of the interviews and workshops we held, we are unable to confirm whether the most important datasets were identified. Nevertheless, geospatial data such as maps, addresses and boundaries were of particular importance in at least half of the ecosystems. In the TfL example, data describing the location of stations is often used to help people access the most relevant transport options nearby. In the Churchill example, administrative geography data is crucial to understand relationships with other open datasets.

Open standards for data

Data standards were used in open data releases across most of the ecosystems. For instance, a data standard was developed following the development of the Leeds Bins app. In the example of local authority spending, two international standards have been central to reducing the additional data preparation and increasing the value of the data released.

Senior-level championing

In many of the examples, we found that systems matured best when the open approach was championed at the senior level, such as a councillor, minister or a senior official, depending on the type of the organisation. These senior leaders highlight the strategic opportunity that the organisation has in releasing open data, prioritise resources, and work to overcome any internal resistance within the organisation. In the Leeds City Council example, the City Council officials' drive and initiative were key success factors leading to articulating the problems, developing solutions, opening up the data, and finally creating the Leeds Bins app.

Intermediaries

Data intermediaries, organisations that have a specific role of ensuring that the available public sector open data is usable, are key actors within many of ecosystems analysed. These intermediaries take data published openly or shared by the public and private sector and, depending on the needs, aggregate and standardise it. It is crucial to acknowledge their role in the process of open data being used to deliver services. The exact form of the intermediary and the activities they undertake depends on the maturity of the particular ecosystem.

Problem focus

Successful integration of open data in service delivery often derives from seeking solutions to a particular problem that the organisation is focused on, rather than publishing open data for the sake of it. In the DWP example, the team developing Churchill took a user-centred design approach to understand how government policy officials used data and then sought out relevant open data to incorporate into the product. Similarly, in the Leeds City Council example, the app came out of an acceleration programme to identify service issues that could be tackled through the use of data and identified a specific problem of citizens not knowing when their bins will be collected. Although alternative approaches might also succeed in identifying how open data can support the delivery of public services, having a problem focus facilitates identifying opportunities. In the examples of Spend Network and the NHS, the focus is on the publication of open data rather than on a problem. This might explain why these ecosystems are less mature in their development than others and why there are fewer use cases.

Open innovation

Early releases of open data seem to have helped push forward and identify further possibilities for improvement. For instance, in the case of TfL, the release of a small number of datasets as feeds eventually led to the development of a more open API system – a development which might not have happened in closed data-sharing agreements. In addition, in most ecosystems, choosing an open approach to data publication attracted more external interest and involvement. For example, more than 14 000 developers are using the TfL data – this would have been difficult if there were access barriers.

Peer networks

In most of the ecosystems, peer networks across organisations emerged to encourage development. In the case of these open data ecosystems, influential networks use open channels and approaches – for instance, connections through Twitter, blogging, and unconferences such as Open Data Camp, LocalGovCamp, etc. This is currently anecdotal, and further research could further examine whether there is clearer evidence that processes and opportunities like these have made a difference in the approach that some of these open data ecosystems have taken.

REFERENCES

Copeland E, Dragicevic N, Simpson H & Symons T (n.d.) Public sector data analytics – A Nesta Guide. https://docs.google.com/presentation/d/1boEBOCNRM8Cx64GA6T ybk21Kyd1QmNuxHY9LLz6DjEY/edit#slide=id.p3

Darabi A (2017) How can policymakers get the most out of data? Ask Churchill. *Apolitical*. https://apolitical.co/solution_article/can-policymakers-get-data-ask-churchill/

Data.gov.uk (2014) Release of data fund update. https://data.gov.uk/blog/release-data-fund-update

Data Mill North – Open Data for the North of England (n.d.) Household waste collections. https://datamillnorth.org/dataset/household-waste-collections

Data Mill North – Open Data for the North of England (n.d.) Leeds bins app record of lookups. https://datamillnorth.org/dataset/leeds-bins-app-record-of-lookups

Davies T (2010) Open data, democracy and public sector reform. Unpublished MSc dissertation, Social Science of the Internet, University of Oxford. http://www.opendataimpacts.net/report/wp-content/uploads/2010/08/How-is-open-government-data-being-used-in-practice.pdf

Deloitte (2017) Assessing the value of TfL's open data and digital partnerships. http://content.tfl.gov.uk/deloitte-report-tfl-open-data.pdf

Department for Work and Pensions – GOV.UK (2017) Churchill, making better use of data https://www.youtube.com/watch?v=hn-Utxqjthg

DWP Digital Blog (2015) DWP's 2020 vision. https://dwpdigital.blog.gov.uk/2015/03/18/dwp-2020-vision-andrew-besford

DWP Digital Blog (2017a) Data for people who don't like data. https://dwpdigital.blog.gov.uk/2017/02/24/data-for-people-who-dont-like-data/

DWP Digital Blog (2017b) DWP Digital: Delivering the government transformation vision. https://dwpdigital.blog.gov.uk/2017/02/17/dwp-digital-delivering-the-government-transformation-vision/

Gov.uk (2013) Shakespeare review of public sector information. https://www.gov.uk/government/publications/shakespeare-review-of-public-sector-information

Gov.uk (2015) Guidance – Local government transparency code 2015. https://www.gov.uk/government/publications/local-government-transparency-code-2015

GovLab (2014) The open data era in health and social care. http://www.thegovlab.org/static/files/publications/nhs-full-report.pdf

HMG (2017) Government transformation strategy. https://www.gov.uk/government/publications/government-transformation-strategy-2017-to-2020/government-transformation-strategy#build-better-tools-processes-and-governance-for-civil-servants

Kimbell L (2015) *Service Innovation Handbook: Action Oriented Creative Thinking Toolkit for Services Organizations*. https://serviceinnovationhandbook.org

Local Government Association (2015) Local transparency guidance. https://www.local.gov.uk/our-support/guidance-and-resources/data-and-transparency/local-transparency-guidance

Nesta (2016) Wise Council: Insights from the cutting edge of data-driven local government. https://www.nesta.org.uk/publications/wise-council-insights-cutting-edge-data-driven-local-government

OECD (n.d.) Rebooting public service delivery – How can open government data help drive innovation? http://www.oecd.org/gov/digital-government/rebooting-public-service-delivery.htm

Open Data Institute (2015a) Environment Agency: Going open. https://theodi.org/ea-going-open-benefits-for-ea

Open Data Institute (2015b) How can open data help improve healthcare? https://theodi. org/blog/how-can-open-data-help-improve-healthcare

Open Data Institute (2017) What is open data and why should we care? https://theodi. org/article/what-is-open-data-and-why-should-we-care/

Open Data Institute (2018) General Transit Feed Specification (GTFS) Case Study. https://docs.google.com/document/d/1m3jJR741VcE6ouyPITWgh6HmASQG2jhfI MCddPAva8U/edit#heading=h.fzurpmmkvtj4

Open Data Institute (n.d.) Open data startups. https://theodi.org/global-network-directory/odi-startups/

Open Data Institute (n.d.) What is data infrastructure? https://theodi.org/what-is-data-infrastructure

Open Government Partnership (n.d.) Stories. https://www.opengovpartnership.org/ stories?country=0&type=0&theme=1196

Prakash M (2017, 1 November) Transformation and innovation – use of data analytics in designing services. Digital Leaders. https://digileaders.com/transformation-innovation-use-data-analytics-designing-services/

Robinson C (2017, 24 November) Putting users first is not the answer to everything. Doteveryone. https://medium.com/doteveryone/putting-users-first-is-not-the-answer-to-everything-dd05b9f11b5

Rogers S (2010, 1 June) Government data: Full text of David Cameron's letter pledging to open up the datasets. The Guardian. https://www.theguardian.com/news/ datablog/2010/jun/01/government-data-david-cameron-letter

Rose M (2010) The complications of a "right to data". MA research project, LLM Intellectual Property, Bournemouth University https://www.slideshare.net/slideshow/ embed_code/key/hqYIWFPoX4yLbf?lipi=urn%3Ali%3Apage%3Ad_flagship3_ profile_view_base_treasury%3B%2BUJNc%2BWdQT6UkBDN9gziRQ%3D%3D

Shadbolt N (2015, 31 May) Open data means business. Open Data Institute. https:// theodi.org/article/open-data-means-business/

Susha I, Janssen M & Verhulst S (2017) Data collaboratives as a new frontier of cross-sector partnerships in the age of open data: Taxonomy development. Proceedings of the 50th Hawaii International Conference on System Sciences. pp. 2691–2700. https:// pdfs.semanticscholar.org/4682/85434d7eb14f0610ffaf6a6f0d591286e9ac.pdf

Transport for London (n.d.) Our open data. https://tfl.gov.uk/info-for/open-data-users/ our-open-data?intcmp=3671

Transport for London (n.d.) Widgets. https://tfl.gov.uk/forms/12425.aspx

Transport for London (n.d.) What we do.https://tfl.gov.uk/corporate/about-tfl/what-we-do

Waugh P (2016, 21 February) Finding the natural motivation for change. pipka.org.http:// pipka.org/2016/02/21/finding-the-natural-motivation-for-change

4.

Localising global commitments: Open data in sub-national contexts in Indonesia and the Philippines

Michael Cañares

The Open Government Partnership (OGP) is a multilateral initiative that aims to secure concrete commitments from governments to promote transparency, empower citizens, fight corruption, and harness new technologies to strengthen governance. OGP's vision is that governments become more transparent, accountable, and responsive to their own citizens, with the goal of improving the quality of governance, as well as the quality of services that citizens receive. Since its inception in 2011, OGP brings together 75 countries and 15 sub-national governments with over 2 500 commitments to make their governments more open. Undoubtedly, the OGP is a global process with nation-states participating in its institutionalisation. In the early days of the OGP, national representatives, who are the pioneer members of the OGP, bound themselves to the ideals of transparency, accountability, citizen participation, and innovation, with the end view of improving governance and public service delivery (OGP 2016). National governments and their civil society counterparts developed national action plans to enact measures to achieve the OGP goals. Sub-national governments became a significant focus of the OGP efforts only in 2015. The move towards the 'local' is conditioned by several normative arguments proposed by the Open Government Sub-national Declaration – that the government is closest to the people at the sub-national level, that the sub-national space offers greater opportunity for transformative change, and that civil society organisations can engage in better discussions with government through sub-national platforms (OGP 2016a). Based on 2015 OGP statistics, a total of 73 out of 1 894 commitments to date address local issues or involve stakeholders from sub-national governments (OGP 2015).

Localising, or the process of localisation of global agenda, has been discussed extensively within the context of environmental protection and development studies. Localisation is a paradigm that favours the local over everything else (in particular globalisation) and results in more control of processes by local actors and communities (Hines 2000). Nevertheless, it is also argued that globalisation and localisation are not competing processes and that the local is a manifestation of greater interconnectedness, hence changes at the local inform global contexts, and global pursuits affect local behaviour (Voisey & O'Riordan 2001). In a recent paper by the Asian Development Bank (2017), it was argued that localising global agenda is not just about integrating global agenda into sub-national plans but empowering sub-national governments to implement those plans and programmes. At the same time, it is about connecting the local with wider national and global debates and how this affects local systems and processes. This reminds us of earlier arguments on globalisation (Escobar 2001), where political processes are institutionalised in one place but are invariably connected to processes beyond it. In the context of open government data, it has been argued that the local is important because, in decentralised contexts, the local is where data is collected and stored, where there is strong feasibility that data will be published, and where data can generate the most impact when used (Cañares & Shekhar 2016). In the case of other global goals, the Sustainable Development Goals (SDGs) in particular, the role of local contexts is seen as critical in transforming global goals to local reality (UN 2014).

At the national level, several building blocks for open government have been identified by scholars. For example, Heeks (2004) argues that external pressure and internal political desire are strong drivers of e-government success and complemented by technology, strategy, design and competency, will ensure the success of e-government. In the case of open governments and writing from an analytical vantage point of transparency and accountability initiatives, Peixoto and Fox (2015) point to the nature of civic engagement as the main driver of downward accountability, and that institutional response is conditioned by willingness and capacity. Recently, in a synthesis paper on open data and sub-national governments across nine cases from six countries, Cañares and Shekhar (2016) argue that there are six facilitating factors for open data initiatives to succeed – political leadership, implementation structure, readily available governance data, technical capacity, existence of intermediaries, and implementation of concrete initiatives. The ADB synthesis paper (2017) agrees with these analyses, emphasising the importance of creating an enabling environment through a legal or policy framework for localisation, financing investments and building the capacity for all actors beyond local governments. This chapter explores to what extent the same characteristics that made open government successful at the national level will resonate in sub-national spaces.

The localisation of OGP is analysed in the context of Indonesia and the Philippines, the pioneering members of the OGP. These governments have launched initiatives to involve citizens in preparing, implementing and monitoring OGP National Action Plans (NAP) since 2012. Furthermore, both countries launched their open data portals as a part of their national action plans – Indonesia in 2014, and the Philippines in 2015. In 2016, Indonesia and the Philippines started to cascade OGP processes and open data initiatives to the sub-national level. This chapter would like to assess the evidence of localising open data initiatives by answering the following questions: (1) what were the drivers for localising open data commitments by local governments; (2) what challenges exist in localising global commitments on open data; (3) what recommendations can be made to enable effective implementation of open data at the sub-national level? Answers to these questions are crucial, especially in a context where the OGP intensifies its efforts to localise OGP processes in several countries, funding initiatives through a dedicated grants window supported by different donors. This chapter is structured in three parts. The first section explains the background and methodology, the second part presents findings and discussion arising from the case studies and the concluding chapter offers recommendations.

Methodology

This chapter analyses two sub-national case studies from Indonesia and the Philippines. Both countries share a similar political trajectory – they experienced authoritarian regimes and are relatively new in terms of democratisation and decentralisation (Fukuoka 2015); nevertheless, there are inherent differences in their local politics (Sidel 2005).

The case study sites of Bohol province in the Philippines and the city of Banda Aceh in Indonesia were chosen from a list of early adopters of open data initiatives. Both Bohol and Banda Aceh are parts of the current sub-national pilot of the OGP action plans of both countries. Moreover, the intention was to investigate different sub-national levels – a province and a city – the former larger in scope and involving several sub-national jurisdictions (e.g. municipalities), and the latter being a single governance entity with a significant population and a myriad of urban concerns. The research was conducted over six months from June 2017 using a primarily qualitative approach. A review of documents, including policies, relevant legislation, project documentation, and prior research papers was conducted at the beginning in order to understand the context of open data initiatives in each place. In each case, a total of 20 key informants were selected for interviews consisting of ten government actors and ten non-state actors. The intention was to gather balanced perspectives from different stakeholders. The narratives from the documents reviewed and the key informant interviews were analysed and common themes were extracted to reveal patterns, areas of agreement, and points of contention. In the analysis, the research considered three factors: (1) the political environment that conditions the open data initiatives; (2) the internal and external pressures that catalyse or prevent the localisation agenda; and (3) the processes that were undertaken to adapt global commitments to the local context.

Findings and discussion

This section will discuss findings from Bohol in the Philippines and Banda Aceh in Indonesia and outline the implications that the findings have on the wider question of localising the open government and open data agenda.

Bohol: Operationalising full disclosure

Bohol is situated in the centre of Visayas region between the south-eastern, Cebu and the south-western, Leyte. The island province of Bohol is the tenth largest island of the Philippines and is composed of one city, 47 municipalities and 1109 barangays. As of 2010, it has a total population of 1 255 128, with 242 307 households with an average household size of 5.1. The main economic sectors are agriculture, industry, fishery and tourism. The Provincial Government of Bohol is one of the stellar performers in terms of local governance and public service delivery because of a professionalised bureaucracy and a thriving civil society. Although provincial leadership is passed from one politician to another in a seemingly scheduled rotation of key political positions (e.g. the governor becomes the district representative in the next elections), the performance of local governance is largely driven by a competent set of civil servants held in regular check by active civil society organisations and local media.

This political landscape enables open government interventions to thrive. When the national government issued the guidelines for the publication of local financial accounts and other relevant information through the Full Disclosure Policy (FDP) portal, Bohol was one of the early adopters, creating an enabling mechanism to ensure compliance. Bohol was rated as fully compliant with the FDP as it published the required information. Data suppliers prepared the required documents for the disclosure and were tasked to submit the documents through a compliance monitor, the Provincial Internal Audit Office (PIAO), which works in coordination with the ICT unit to ensure that all data requirements are uploaded. Data uploading was done by the ICT unit whose main task was to look after the government IT systems and the website. Uploading documents to the website and the FDP portal is just a fraction of their total mandate.

However, although a compliance mechanism was set up inside of the local government, convening different agencies, systematising a timeline for data availability and disclosure, and assigning a central compliance agency, there was no initiative to inform and build the capacity of the users. The FDP is a part of the enabling mechanisms of a national commitment from the first OGP national action plan in the Philippines and has been reported as one of the important actions that contributed to the completion of the commitment (Mangahas 2014). The system has been cascaded to the local level to support transparency and information disclosure; nevertheless, no effort has been made to inform the public about the existence of the portal or its use. The good intentions of openness fell short of enabling participation (Cañares 2014).

An action research project funded by the International Development Research Centre and implemented by Step Up Consulting Services aimed to enhance citizen engagement through open government data and was implemented in 2015 to 2016 in Bohol and the neighbouring province of Dumaguete. The project was the first time when stakeholders in Bohol (advocacy groups, non-government organisations, and people's collectives), were trained on how to access the data from the portal, scrape the data (from PDF), convert it to a format that would allow analysis (e.g. CSV, or comma-separated values), and use the data and its corresponding analysis to ask government certain questions, demand explanation, and initiate dialogue on key issues of their interest (Cañares et al. 2016).

Before the training, only one out of the 21 organisations surveyed had accessed and seen documents uploaded onto the FDP portal. The majority of them accessed government data through formal (e.g. writing a letter) or informal means (e.g. asking friends from within the government) even when the data that they were requesting was already available on the portal, simply because they were not aware that the portal existed (ibid.). For the local government, their task was to make information available; they did not perceive engaging with users and increasing their understanding of technical documents as a part of that process. However, this is a result of the programme design rather than the implementation in Bohol. The national government agency that launched the initiative (Department of Interior and Local Government (DILG)), designed the FDP primarily as a data and information disclosure mechanism, without closing the feedback loop. Hence, the objective of the initiative is largely intended to make the data available, not necessarily to encourage citizens to engage with the government. Nevertheless, even at the DILG level, there has been a lack of initiative to educate users, and this was generally true for the rest of the country at the time when FDP was implemented. Most of the data literacy trainings – if and when they happened, similarly to the case of Bohol – were initiated by actors outside of the government. The capacity building training launched in Bohol by Step Up Consulting resulted in a number of positive outcomes. Regular mentoring activities increased the knowledge and skills of the participants to use data in their advocacy and development work, including advocating for better use of the local government's disaster risk reduction and management funds. This was particularly relevant as the province suffered a major environmental disaster shortly before the training was conducted – a 7.2 magnitude earthquake that resulted in damage to the local economy worth USD 47 million.

Banda Aceh: From Freedom of Information to open data

Banda Aceh is the capital and the largest city in the province of Aceh, located in the north-western tip of Indonesia. Home to approximately 248 727 people, the majority of whom are Muslim, the city is recovering from a devastating tsunami that claimed lives and property approximately ten years ago. The city has been the centre of conflict for many years, in recent times between the Indonesian national government and secessionist local forces. Nevertheless, the post-tsunami mobilisation seemed to improve national and local alliances and the city has been relatively peaceful since. To date, Banda Aceh enjoys a level of autonomy, implementing Sharia law strictly within its boundaries.

In 2008, the Indonesian government introduced the national Freedom of Information Act (FOIA), providing legal guarantee for citizens on how to access information held by government institutions at the national, regional and district levels through various request channels. The city government of Banda Aceh introduced a wide range of policy measures to promote unrestricted access to information at the city and sub-district levels. The government of Banda Aceh is considered as one of the most responsive to citizen information requests – the Information and Documentation Management Officers (PPID) ensure that the information provided by authorities is not only compliant with the law, but also satisfactorily fulfils the needs of citizens. The city of Banda Aceh ranked first out of 23 municipalities and districts at the provincial FOI Awards in 2014[1] and in 2015.[2] Despite this recognition, the uptake of FOI in Banda Aceh is considered as low. PPIDs only received a total of 40 FOI requests, submitted either online or offline throughout 2014. They responded to 34 FOI requests and provided citizens with information such as school profiles, teacher profiles, population data, and the results of the general election. A total of six FOI requests were denied due to the data not being available. The figures for 2015 were similar – the city government received 36 requests and almost all were resolved.

The FOI/OD Banda Aceh project was introduced in 2014 with the aim of minimising the information gaps between the government and citizens by harnessing the potential of open government data. Implemented by the World Wide Web Foundation's Open Data Lab Jakarta and funded by USAID under its Kinerja project from October to December 2014, the FOI/OD Banda Aceh project complemented the existing FOI efforts to increase the supply, demand and use of public information. Thus, the project encouraged the city government to practise proactive data disclosure, while simultaneously building capacities of Civil Society Organisations (CSOs) to make meaningful use of this data. The initiative had high political support with a desire to demonstrate its progression towards democracy embracing the information technology.

---

1   http://perhubungan.bandaacehkota.go.id/v3/pemko-banda-aceh-terima-anugerah-kip-2014/
2   http://www.lintasnasional.com/2015/12/10/komisi-informasi-aceh-gelar-anugerah-keterbukaan-informasi-publik/

The first phase of the project supported the education agency in disclosing 17 datasets requested by CSOs. Some of these datasets included school profiles, teacher profiles, school budget, and online student admission. The project resulted in the launch and operationalisation of the Banda Aceh open data portal[3] with two more departments participating in proactive disclosure other than the education agency (e.g. transportation and communication agencies). The project trained 27 civil servants on open data and proactive disclosure and 25 representatives from 19 CSOs in accessing, understanding, and using the open data published by the education agency. This process was replicated for the health and transportation datasets. The number of data downloads from the portal exceeded the number of FOIA requests by more than 700. The ease of accessing government data through the portal in comparison to filing paper requests was a significant factor contributing to its popularity.

However, only a few departments participated in the open data portal. In the meeting with Dishubkominfo, the Badan Perencana Pembangunan Daerah (Bappeda) planning agency, and the Sekda (city secretary), the city mayor of Banda Aceh expressed the need to institutionalise proactive disclosure in the city through a mayor's legislation and a concrete action plan. This proposal was seconded by CSOs headed by Gerak Aceh, an anti-corruption non-governmental organisation. An initial discussion of the draft regulation was completed in the early part of 2016 and promulgated into a mayor's regulation before the mayor stepped down from her post, having formulated a roadmap for proactive information disclosure in the city.

Similar to the Bohol case, national imperatives, alongside demands for better access to data from civil society groups, facilitated the data disclosure process and made it more efficient. What is unique about the Banda Aceh case study is the strong buy-in from the information and communication agency in pushing for the data disclosure agenda that culminated in the passage of local policies to support the requirement for data disclosure from other government agencies.

From global to local: drivers and enablers of data disclosure

What were the drivers that enabled the process of localising the open government and open data agenda? Looking at the case studies of Bohol and Banda Aceh, the following key ingredients emerge: leadership, civil society, legislation, data literacy and approach.

First, political commitment is key. Sub-national spaces led by government leaders who perceive transparency and accountability as important ingredients in improving public service delivery and have the buy-in of key department officials from critical agencies have a higher likelihood of success.

---

3   http://data.bandaacehkota.go.id/

In Banda Aceh, the city mayor and the officials from the planning agency and the information, communication and technology office were the main initiators in pushing for greater transparency through open data. However, scaling the pilot interventions that were initially started in two departments was met by considerable resistance from several government officials who were looking for a legal basis in the practice of open data. Without an enabling national law on open data, local partners drafted a mayor's regulation on data management that contained provisions on open data and proactive disclosure. Without the regulation, increasing the number of departments covered and the number of datasets published would have been an arduous task. Surely, this regulation would not have been possible without the strong push from the departments, and the commitment of the city mayor who issued them.

Second, a vibrant civil society is critical in pushing for and sustaining open government initiatives, including open data. Without civil society participation, such as advocates, watchdogs, initiators for better transparency, accountability, and openness, open government and open data processes might become supply-led and irrelevant. Therefore, civil society should have the capacity to engage proactively with the government. The relative power of civil society to exert influence on local government leadership is crucial. In the case of Bohol, civil society was considered as a governance partner, largely brought about by policy requirements that mandate civil society engagement and collaboration on key development issues, especially during the time when Bohol was considered one of the 20 poorest provinces in the country. In Banda Aceh, the relationship was initially more antagonistic, as the primary mover of the transparency initiative from the civil society side was an anti-corruption NGO. Open Government Partnership and its principle of co-creation require a thriving civil society. However, it might not be the case in most sub-national governments. For instance, in the Philippines, while procurement legislation requires that contracting activities are observed by civil society organisations, in some jurisdictions, there are no qualified civil society observers. Here, co-creating government plans may not be feasible. The reason why it was successful in both, Bohol and Banda Aceh, was the long tradition of civil society participation in local governance, conditioned by the presence of local actors working on specific issues and interacting regularly with the government. This is not necessarily true across the whole of Indonesia and the Philippines.

Third, building local capacity within the government and civil society groups is key. In the case of open data, this is where data literacy that shows concrete results is fundamental in using open data to promote better governance. Raising awareness of the importance of proactive disclosure mechanisms and inciting stakeholders' appreciation of the power of data is an important initial step. Once the value of data is clear to data suppliers and users, engaging them in skill-building activities will be easier and faster. However, skill-building activities should be able to demonstrate concrete results to sustain interest in data-driven approaches. What the data reveals is important, but what can be done with it is critical in generating the value out of open data. Often, data trainings end with generating insights and revealing information that was unavailable without access to datasets. Thus, data trainings should lead to concrete actions that governments and citizens can act on.

Finally, open government and open data can work if the initiative is latched onto problem-solving exercises that push for better delivery of public services. Problem-driven data approaches cultivate and incentivise data use. Engaging the users to begin identifying the problem that different stakeholders are passionate about contributes to more effective implementation of data-driven solutions. In most cases, these are problems that affect people's daily lives, such as a high malnutrition rate in children, or finding a locally made product to promote and sell. Having identified the problem to solve, we ask the following questions: what data do we need to be able to find solutions to these problems and how can we get it? This approach helps ensure that when the requested government data will be made available, it will be accessed and used. It also ensures the government of the importance of disclosing the data for solving problems that governments ought to address.

Creating a 'culture of data' requires a conscious choice to base decisions on data only; this includes analysing the political implications of disclosing or not disclosing data. Data openness levels the playing field in terms of exerting power in policy-making as the same datasets can become the basis for contestation.

Challenges in localising the global agenda

Localising the global agenda on open data has many challenges. The case studies in this chapter demonstrated the following challenges: the lack of consultations with sub-national governments where national commitments were made; lack of a legal and institutional framework to guide the efforts for localisation; unrealistic expectations of the quality of civil society organisations at the sub-national level; and political transitions affecting the sustainability of reforms.

First, global commitments on open government and open data are discussed at the global and national levels, without necessarily conducting wider consultations with sub-national governments. The Open Government Partnership is an initiative of national governments that has been cascaded to sub-national platforms only recently. The awareness that sub-national leaders have of these commitments is uneven, with more proactive leaders and those with strong participation in national discussions being relatively more aware. In both Bohol and Banda Aceh, the open data awareness of local leaders was externally driven by researchers and international organisations. The receptiveness of local leaders to this external push was critical for the agenda to succeed locally.

Another barrier is the lack of a legal and institutional framework to localise global commitments. In Indonesia, open data initiatives are hampered by the absence of a national legal framework that requires proactive data disclosure in open formats. In the Philippines, there is specific operational guidance on how local governments undertake data disclosure, yet it does not contain open data provisions. Furthermore, there is a lack of policy mechanisms where local governments can contribute to open government besides the Full Disclosure Policy requirement. Given that there is no 'localisation' design, most of these initiatives when cascaded to sub-national level depend on the strong political commitment of local leaders, the creativity of the local bureaucracy, or the push of civil society organisations.

Third, political transitions pose a significant risk to the sustainability of open government reforms. To ensure that open data reforms are immune to political transition remains a challenge, especially when these are not institutionalised. In Bohol, this problem was mitigated due to the re-election of the previous governor for his third term. However, due to the three-term limit, the transition will occur shortly. In Banda Aceh, several stakeholders were apprehensive because of the upcoming elections during the key phase of implementing open data initiatives. In a strategic move to preserve open data practices in the city, different stakeholders from inside and outside of the government pushed for the issuance of a mayor's regulation. This was a critical move as the mayor who supported the initiative and issued the regulation lost the elections. Nevertheless, his open data regulation is still in effect.

Finally, technology-enabled interventions require investments in skills, infrastructure, and enabling conditions. In decentralised contexts where local revenue is insufficient, this may be deprioritised in favour of social and economic programmes, besides the fact that these may be beyond the capacity of poor, small local governments to invest in. Hence, as indicated in the cases of Bohol and Banda Aceh, open data initiatives happened because of having significant external donor support. In Banda Aceh, donor support enabled investment in capacity building for suppliers and users of data. In Bohol, local government capacity was already high, partially owing to donor-funded governance programmes focused on building bureaucratic capacity. Nevertheless, investment in improving civil society capacity to work with data was provided by external donors.

Conclusion

For open data to result in more transparent and accountable governance at the sub-national level, there is a need to ensure that the local governments are able to disclose key information about how the government is carrying out its mandate, and that citizens have the capacity to engage and use this information to scrutinise government functioning, help identify solutions to local problems, and promote a conversation to close the feedback loop in service delivery. Developing a process that brings together different stakeholders to define and agree on common goals and use data is a crucial investment in time and resources.

The results of this research are important, especially because the Open Government Partnership has been extending the implementation of national pilots of the OGP process, including the open data pilots. This chapter could help inform how sub-national pilots are structured, implemented or monitored. Nevertheless, the results of this research are limited only to the cases covered in this study. Future investigation is needed to analyse cases in different countries, particularly those where the open data agenda is largely defined by national governments.

This chapter offers the following key recommendations in terms of localising open government and open data commitments:

a. Engage in local consultations with local government units. National commitments that affect local governments need to be a product of consultations with local leaders. National governments need to improve communication with the local leaders as high-level political support is key in advancing open data initiatives. Incentives to participate in open data initiatives need to be clearly communicated to the local leaders; for instance, rewards for excelling in the initiative or finding solutions to long-standing problems faced by citizens and their communities.

b. Ensure foundational legal basis for proactive data disclosure and the implementation of open data initiatives. The absence of legal basis poses a significant challenge for local governments to exact commitment from local agencies. Having a legal framework can institutionalise open data initiatives and insulate these against changes in leadership.

c. Provide avenues for citizen engagement through selected representatives (with civil society organisations acting as intermediaries).

d. Invest in local capacity for data suppliers and data users. The focus of capacity development initiatives depends on the initial needs assessment. In some cases, this will require providing a basic technical infrastructure along with technical training. In others, it will work to improve the ability of users to access and use the disclosed government data.

REFERENCES

Asian Development Bank (2017) *Localizing Global Agendas Report on the Joint Learning event of the Governance Thematic Group and the Development Partners Network on Decentralization and Local Governance*. Manila: ADB

Cañares M (2014) Opening the local: Full disclosure policy and its impact on local governments in the Philippines. Paper presented at ICEGOV '14: Proceedings of the 8th International Conference on Theory and Practice of Electronic Governance, Guimaraes, Portugal, October 2014. pp. 89–98. http://dl.acm.org/citation.cfm?id=2691214

Cañares M, Marcial D & Narca M (2016) Enhancing citizen engagement with open government data: The case of local governments in the Philippines. *The Journal of Community Informatics* 12(2): 69–98. http://www.ci-journal.net/index.php/ciej/article/view/1256/1208

Cañares M & Shekhar S (2016) Open data and sub-national governments: Lessons from developing countries. *The Journal of Community Informatics* 12(2): 99–119. http://www.ci-journal.net/index.php/ciej/article/view/1260/1209

Escobar A (2001) Culture sits in places: Reflections on globalism and subaltern strategies of localization. *Political Geography* 20(2): 139–174

Fukuoka Y (2015) Who brought down the dictator? A critical reassessment of so-called 'people power' revolutions in the Philippines and Indonesia. *The Pacific Review* 28(3): 411–433. DOI: 10.1080/09512748.2015.1011212

Gaventa J (2006) Finding the spaces for change: A power analysis. *IDS Bulletin* 37(6): 23–33

Heeks R (2004) Causes of e-government success and failure: Factor model. Institute for Development Policy and Management. Manchester: University of Manchester

Hines C (2000) *Localization: A Global Manifesto*. London, New York: Earthscan

Mangahas M (2014) *Independent Reporting Mechanism: Philippines Progress Report 2011-2013*. Open Government Partnership and the Philippine Center for Investigative Journalism. https://www.opengovpartnership.org/wp-content/uploads/2019/07/IRMReport_Phillipines_100813c.pdf

Open Government Partnership (2015) Subnational governments and the open government partnership: Issues and Options Paper. Unpublished document.

Open Government Partnership (2016a) Open by default, policy by the people, accountability for results. Washington, DC: OGP

Open Government Partnership (2016b) Open Government Subnational Declaration: Paris-France 2016. Washington, DC: OGP

Peixoto T & Fox J (2016) When does ICT-enabled citizen voice lead to government responsiveness? World Development Report. Unpublished Background Paper. http://hdl.handle.net/10986/23650

Sidel J (2005) Bossism and democracy in the Philippines, Thailand, and Indonesia: Towards an alternative framework for the study of 'local strongmen'. In: J Harriss, K Stokke & O Törnquist (eds) *Politicising Democracy: The New Local Politics of Democratisation*. UK: Palgrave Macmillan. pp. 51–74

United Nations (2014) *The Road to Dignity by 2030: Ending Poverty, transforming all lives and protecting the planet*. Synthesis Report of the Secretary General on the Post-2015 Agenda. http://www.un.org/disabilities/documents/reports/SG_Synthesis_Report_Road_to_Dignity_by_2030.pdf

Voisey H & O'Riordan T (2001) Globalization and localization. In: T O'Riordan (ed.) *Globalism, Localism, and Identity: Fresh perspectives on the transition to sustainability*. London, New York: Earthscan. pp.25–42

5.

Closing the gaps in open data implementation at sub-national government level in Indonesia

Ilham Cendekia Srimarga & Markus Christian

Given the rapid development of information technology for the past fifteen years, there are people in various quarters who believe that this has spurred wider adoption of an open ICT ecosystem.[1] Such an ecosystem is believed to be instrumental in transforming (especially) developing countries in reaching their development goals.[2] With that belief, many governments (national and local) have opted to adopt open ICT ecosystems to overcome their development challenges.

Nevertheless, there have been questions whether ICT projects have the intended impact. Research was conducted 'to understand the impact of open data projects in terms of changes in the relationship between an intermediary organisation and a data supplier (government), and between the intermediary and the end-user' (Maail 2017: 153). The study presented in this chapter is of a kind that applies the action research approach (direct involvement of the researcher-participant in changing the research subject) by looking at the process of open data conception and implementation in a district and how the response dynamics took shape.

The approach considered the combination of the local specific needs and the generic and replicable data design to show that there is a significant mutual shaping between local needs and the ICT design put forward by the technology designers (see Orlikowski 1992; Diniz et al. 2014).

---

1   'An ICT ecosystem encompasses the policies, strategies, processes, information, technologies, applications and stakeholders that together make up a technology environment for a country, government or an enterprise' (Open ePolicy Group 2005: 3).
2   Smith M & Laurent E (2010) Open ICT ecosystems transforming the developing world. *Information Technologies and International Development* 6 (1): 65–71.

The study was carried out in Bojonegoro, Indonesia, a district that actively builds innovation, develops open government and promotes good governance, particularly since 2014. ICT utilisation for addressing development problems is one of the policy priorities of Bojonegoro district government. At that time, the regent (or mayor) of Bojonegoro district, Suyoto, believed that open and innovative government is a road the district must take to achieve its local development goals, particularly as part of its improvement in education and health as well as poverty eradication. He also believed that open government means involving the community to reach the development targets.

One of government's openness strategies is open data. In this context, open data was translated as the provision of data that is open to all actors, both state and non-state actors (such as NGOs, CSOs, community-based and faith-based organisations). Open data is expected to allow state and non-state actors to access and immediately use the data to contribute to achieving the local development goals of Bojonegoro. Some of the main goals of local development are, for example, reducing poverty, increasing enrolment rate, improving access to health services, and improving infrastructure.

The government of Bojonegoro implemented open data to facilitate its open policy. The open data implementation has two phases. In the first phase, the district government implemented it single-handedly. Some activities were (1) opening up public data to the citizens in the district; (2) collaborating with the central government, that is the UKP4[3] (now presidential staff office or KSP), to implement One Data system; and (3) conducting public dialogues regularly in which data is opened up to the participants (the public/citizens).

The Bojonegoro government utilised its web portal as the channel for opening up the data to the public. However, the data was still in aggregated format and mostly not up to date. The data displayed was not comprehensive and in a format that was difficult to process. On the other hand, the demand for data was still low. The citizens and communities still failed to comprehend the data and did not have the capacity to use and benefit from the opened data.

Since the results of the first phase of open data implementation were considered as below expectation, the Bojonegoro government decided to introduce improvement. They took the second phase of open data implementation in which the government welcomed cooperation with various stakeholders, especially CSOs. These CSOs were expected to provide analyses and recommendations on the open data implementation in the district. The space given by the government motivated Sinergantara to get involved by introducing the idea of 'data revolution' in Bojonegoro.

---

3   UKP4 stands for Unit Kerja Presiden Bidang Pengawasan dan Pengendalian Pembangunan, set up primarily to monitor ministries and central government agencies' works and to ensure acceleration of government programmes. It was formed under the administration of President Susilo Bambang Yudhoyono (2009-2014) and now replaced by KSP (Kantor Staf Presiden).

The involvement of Sinergantara and other CSOs in identifying the problems, conceiving the solution and implementing it, as well as the temporary changes produced by the implemented solution, is the subject of the action research presented in this chapter.

The research questions posed by the research are as follows:

1. What are the factors or issues that may lead to disconnections[4] (gaps) in open data implementation at the sub-national level (in this case, in Bojonegoro district, Indonesia), in relation to the government's expectations on the contribution of open data to efforts of achieving local development goals?
2. What are the 'closing the gaps' solutions developed by Bojonegoro district government, and what are the processes and background that led to the solution? How does collaboration of sub-national governments and non-state actors contribute to the birth of the solution?
3. How does the implementation of the solution help improve effectiveness of open data at the sub-national level? What has been changed, particularly among the actors involved in the open data implementation? (Changes at target groups/beneficiaries which may not be visible at the moment.)
4. What are lessons that can be drawn from the experience of open data implementation in Bojonegoro district (from identification of disconnection/gaps to implementation of solutions for closing the gaps) for knowledge on open data implementation at the sub-national level in other areas?

The objectives of the research are

1. To describe the lessons drawn from the experience of Bojonegoro district government (sub-national government) in Indonesia with regard to:
   a. identification of obstacles that led to disconnection/problems in their open data initiative implementation;
   b. development of solutions that allow the realisation of open data to support the efforts of achieving their local development goals;
   c. utilisation of disaggregated data (single data) for solving problems that usually create gaps in open data initiative implementation.

---

4   In our understanding, there are several types of disconnection: (1) disconnection due to physical distance; (2) disconnection due to nonexistent communication; (3) disconnection due to limited capacity of the beneficiaries; (4) disconnection resulting from complexity of reference/information coming from the supply-side (open data provider); (5) disconnection due to management aspects; (6) disconnection due to limited (or unavailable) space for participation; and (7) disconnection due to unavailable information. The issues raised in this paper are connected to points 3, 4, 5.

2. To explain possible impacts (or potential impacts expected to happen in the near future) as a result of the innovation in Bojonegoro district. This chapter will look at these impacts from two sides: (1) benefit of open data innovation to the efforts of achieving local development targets; (2) contribution of this solution of innovation to the maturity of open data approaches at sub-national government level.

Methods

Research that underlies this chapter is action research, in which the researcher (author) is involved in the interventions or efforts of making this initiative work during implementation. The author of the chapter acts as a researcher-participant, getting involved in the process of providing solution(s) as well as the implementation (Kemmis, McTaggart & Nixon 2014; McNiff & Whitehead 2002).

The researcher took four steps in the research: (a) assessment; (b) intervention; (c) observation; and (d) reflection. These steps are modified from the standard steps of action research.[5]

Assessment

Several steps were taken as part of the assessment, which aimed to review the existing situation in order to develop the next steps of the action research:

- Understanding the ecosystem of the open data initiative implementation in Bojonegoro district. By the ecosystem of open data, we mean a set of relationships consisting of actors, relations among actors, artefacts, relation between actor and artefact and how the situation influences open data initiative implementation (whether it supports, inhibits, or distracts).
- Identifying the business processes of the ecosystem. In this case, we see what business processes (activities) occur within the open data implementation and the progress they make (smoothness of the activity's workflow).
- Understanding critical points that result in obstacles inhibiting the activity flow. On these points, obstacles came from various factors leading to the situation that we call disconnection (gap).
- Understanding of the gap as illustrated by the visual scheme which makes people understand which points suffer gap/disconnection.

---

5   Most action research consists of standard cyclical steps comprised of: (a) planning; (b) action; (c) observing; (d) reflecting (see McNiff & Whitehead 2002). As there is the element of participation of the researcher, we also refer to Kindon et al. (2007).

Intervention

After identifying the situation, the researcher devised interventions to help solve the problem. Basically, the interventions consisted of actions to identify the gaps of the first open data implementation in Bojonegoro and actions to develop a solution to address the problem.

Several activities were conducted to enable the author to gain insight into the gaps and or problems affecting the first process of open data implementation in Bojonegoro: (1) Interviews with various stakeholders on existing issues of open data implementation in the district. The interviews were in the form of focus group discussions (FGDs) and face-to-face meetings. (2) Close engagement with the then regent of Bojonegoro, Mr Suyoto (also well known as Kang Yoto). Kang Yoto expressed his strong interest in developing and adopting open data in his district and shared many of his ideas about the issue. It was the regent who provided support to the development and implementation of the solution.

To complement the above activities, the researcher also carried out desk study and consultation with the staff of the government of Bojonegoro concerning the existing and yet-to-be developed regulation on open data implementation in the district. The researcher also studied the technological system (including technical elements) of the open data in Bojonegoro as employed in the first phase of the programme. Critical focus is directed toward the way data is collected, stored and processed.

Since open data requires participation and the solution must also involve this aspect, the researcher then engaged with local CSOs to get insight about the view and perspective of the demand side regarding the open data implementation in the district.

Having all the needed information and insight on the obstacles and other inhibiting issues of the first process of open data implementation in Bojonegoro, the author and his organisation conducted solution-related intervention.

The first was the inventory and identification of the existing solutions (that have been developed) by state and/or non-state actors. The researcher looked into what has been done to address the issues and bridge the gaps. Afterward, the researcher conducted further elaboration by seeking agreement from the government and CSOs on what kind of solution the district needed.

Those two activities became the basis of developing a solution in which the researcher and his organisation conducted the following:

- Developing 'data revolution' methodology consisting of: (i) development of manual on data revolution methodology; (ii) development of Android-based Revo-application installed on smartphones and desktop-based dashboard to store and process the data collected from the application.
- Conducting participatory data collection using the application. The data was stored and processed in the desktop. Some volunteers in certain

91

selected villages in Bojonegoro applied the methodology by using the app and allowed the collection of basic data to be stored in the system.

- Providing training and short education courses regarding the data revolution methodology by involving government staff and volunteers from the villages.

Observation

Parallel with and after the intervention, the researcher conducted observations. As the study was an instance of participatory action research, the researcher had been actively engaged with other stakeholders and directly involved with the implementation of the solution that he and his organisation devised. At the same time, the researcher took a distance and looked at the ongoing process from the 'outside'. This was intended as a means to monitor and record the process in order to generate the data.

Two issues became the framework for conducting observation:

- The process of translating the interest of the regent into a technological system (open data system): The researcher came forward with the solution to the problem of the first process of open data implementation by developing a technological system (application and dashboard). The question was how this technology was received by and the response of the stakeholders (supplier of data within the government and the users among the citizens/CSOs).
- The second open data implementation developed a new kind of data, one of which is the social data consisting of individual and household data representing the timely condition of each individual and family as well as certain conditions (like high-risk pregnant mothers and those living in poverty). How has the social data been received, responded to, and adopted by the stakeholders?

Reflection

Reflection was the next step following action and observation in action research. Reflection is based on the data collected during the action phase in which the author conducted various activities to introduce the solution to the existing situation. Reflection is also a backward-forward look in action research. It is backward as it looks at the action already taken, while it is also a look forward since the result of the reflection will become the basis for the next action research (or just action).

The focus of the author of this chapter in reflection has been on the possible change that already happened due to the introduction of the solution (data revolution) to the open data implementation in Bojonegoro. Several issues were reflected on based on the data collected during the action:

a. Impacts of effectiveness of implementation. The author looked at the following aspects and reflected upon the changes that may occur:

- Sub-national government policies related to this initiative.
- Sub-national government's work plan related to implementation of this initiative.
- Budget availability of sub-national government and CSOs related to implementation of this initiative.
- Availability of work team from government side and CSO to support implementation of this initiative.
- Progress of (work) implementation of the initiative.

b. The closing of the disconnection by the implemented solution. By looking at the problems related to the disconnection; whether they had been addressed or not by the solution. Here, the author reflected whether the solution already addressed the following problems:

- Late update of available data.
- Difficulty in processing the open data[6] provided.
- Difficulty in verification of the open data provided.
- Inconsistency between open data that is provided and data that is needed.

Findings

This research began with the assessment phase in which the researcher identified the issues and possible problems in the subject of the research, which was the first phase of open data implementation (or the first process) in Bojonegoro district. It was found that the district government had a certain assumption about open data, and it led to the way the implementation was carried out.

Figure 1 below explains what the regent of Bojonegoro expected from open data and the collaboration of state and non-state actors.

First, it was assumed by the regent and the staff of Bojonegoro district that the existence of open data allows state and non-state actors to find and have better information sources, so that they can contribute more actively in achieving local development goals. Second, they assumed that the actors have developed the habit (familiarity) and ability of using or processing data (in all formats) into specific information that can be immediately used to contribute to achieving local development goals. Third, it was assumed that in terms of quality, open data has certain sufficiency and comprehensiveness that allows them to be processed into the desired information.

---

6   Here, 'open data' is understood as data already integrated into open data system, and digitised version of GDCS (Gerakan Desa Cerdas dan Sehat / Healthy and Smart Village Movement) data. This GDCS data was collected manually by village volunteers in Bojonegoro.

Figure 1. Expectation and assumption of how open data initiative is implemented

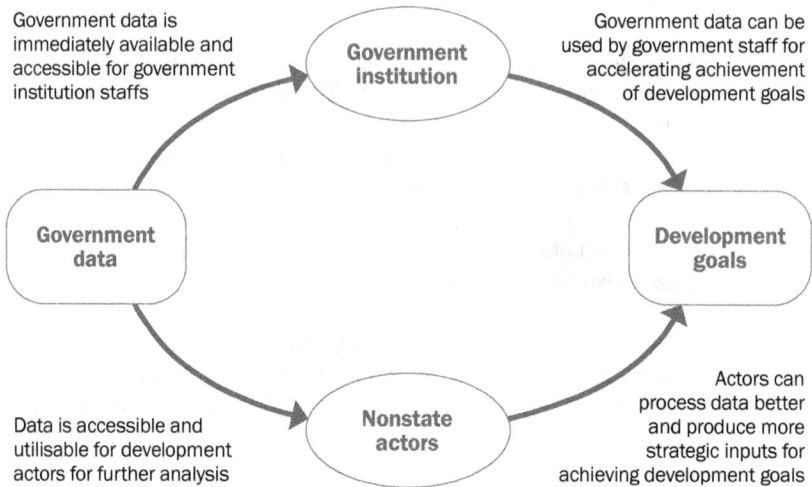

These assumptions were based on the knowledge of the regent of the importance of open data for development. Especially in regard to the first assumption, the regent was inspired by the thinking of Jeffrey Sachs (2015 a, b). Data is considered as vital to development and for data to be instrumental it should be open and accessible. The regent's thinking was also sparked by the newly implemented (at the time) freedom of public information act (Undang-undang Keterbukaan Informasi Publik Nomor 14 Tahun 2008).[7] In its implementation, the law only required the government to publish its information to make it accessible to the public. With that, the government was considered as innovative without having seen the real impact of the implementation to local development.

However, during the implementation of open data, some obstacles (disconnection or gap between what is expected and what actually happened) were found that made the utilisation process of open data fall short of expectations. Figure 2 visualises the obstacles found. The obstacles are related to quality of data and its attendant consequences.

The lack of quality of the open data makes state and non-state actors find it difficult to use and process the published data (data made accessible to the public) into more significant information that can contribute to the efforts of achieving local development goals. We conform to the definition of data quality as 'the degree to which a set of characteristics of data fulfils requirements. Examples of characteristics are: completeness, validity, accuracy, consistency, availability and timeliness. Requirements are defined as the need or expectation that is stated,

---

7   The law was passed in 2008 and it took effect in 2010.

94

generally implied or obligatory'.[8] For example, the first process of open data provided only the manual data which was then published on the web, bringing with it the characteristic of the 'old' data, which was aggregated, not up to date, thus inaccurate and not valid (as it was too obsolete and incomplete).[9]

Consequently, the lack of data quality led to less than effective openness; local actors have access to data but still find it challenging to really use the data. It was further prohibited by the discrepancy between available data (data provided) and their need for data. Thus, data utilisation by local actors in supporting local development goal achievement is far below expectation. Unfamiliarity with the data among the actors and the lack of quality of the data together make it difficult for the actors to contribute to the efforts of achieving local development goals.

Figure 2. Diagram of problems in open data initiative implementation

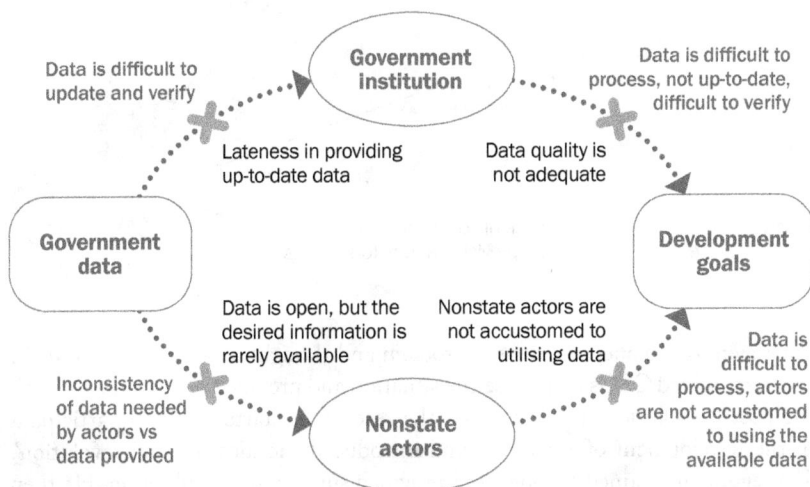

Put simply, problems related to the open data initiative in Bojonegoro district are low quality of data and low capacity (or familiarity) of local actors in using the data effectively. After a series of FGD activities with stakeholders, we found factors contributing to both problems were data update and data verification processes that do not occur immediately (or usually lagging) in the district. Also, the available data (which is in aggregated format only) is difficult to process as

---

8   This definition from Wikipedia is just one of many definitions and it is related to the ISO 9000 definition of quality. As to the issues of data quality, see Redman TC (2008) *Data Driven: Profiting from Your Most Important Business Asset.* Boston: Harvard Business Press. p. 41.

9   Data on schools, for example, only shows the aggregate number without providing details of each school (the single data). The data is the result of a survey conducted one year past at the latest and/or even earlier than that (two to three years past).

well as it does not suit the needs of local actors. The root of these problems is the unavailability of disaggregated data.

Some of the main reasons for the problems are illustrated in Figure 3.

Figure 3. Logic structure of root of open data innovation problem in district government

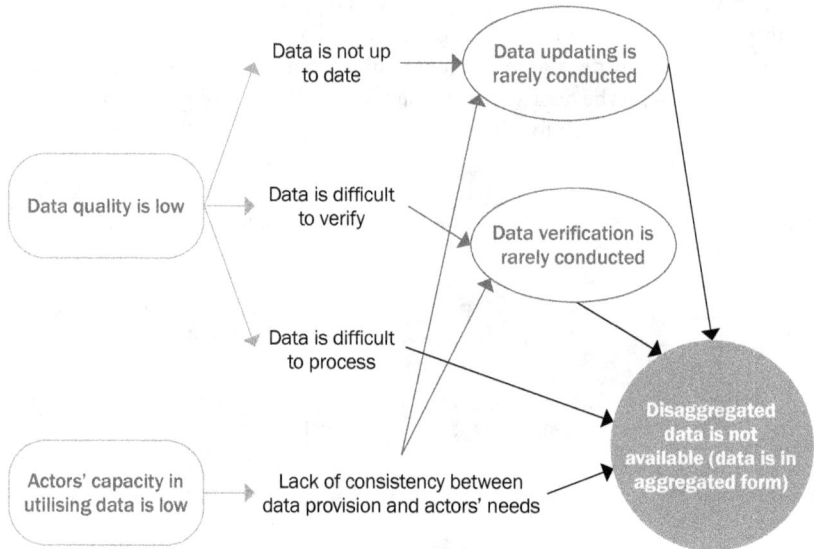

The regent of Bojonegoro saw the problem and thought of ways to overcome it. He then invited CSOs to analyse the situation and provide the government with the possible solution. It is here that the researcher started to be a participant in the development of a solution and introduced the idea of 'data revolution'. The regent entertained the idea and provided support for its adoption. He then decided to initiate data revolution which contains systemic efforts of improving data quality, availability, and update. The data revolution is implemented through, among many efforts, providing disaggregated data (or single data) that is also called data at zero level. Data at that level can be used to monitor the situation of the population of the district, whether it is the individual or the family. Using data at zero level, data update and verification processes can be conducted more easily and quickly.

The regent of Bojonegoro thought the idea would suit the situation of Bojonegoro district at that time. The district government had already had a data collection routine which is conducted regularly throughout the whole year. The activity, called 'Dasa Wisma' data collection (data per 10 households, collected by each neighbourhood), has produced data per individual citizen and per family of each village which is updated every 6 months. The data is in disaggregated form. By applying aggregation and analysis, the district government considered the data as instrumental in measuring development achievements and providing early warning. However, until 2016, the data was available only in manual/ non-digital format, and the structure still needed improvement to allow it to be processed further.

In 2016, data revolution[10] and open data contract initiative was included as part of the sub-national Open Government Initiative (OGP) of Bojonegoro district government.[11] This was like the second launching of open data for Bojonegoro government, after the open data initiative in 2014.

Following the second launching of open data in September 2017, there were several steps that the researcher-participant took. The first was the development of technology, especially for the purpose of data structuring and digitising. The researcher-participant created a data revolution application aimed at ensuring the quality, availability and update of data. The second was the involvement of the citizens from the very beginning of the process through their active participation in single data (disaggregated data) collection. This participation allowed the citizen to have a first-hand experience of being involved in developing disaggregated data. It also resulted in their understanding of the benefit of disaggregated data and the usability of such data for local development.

Both steps resulted in a data format that suits the data revolution approach: well-structured and processable digital data which is regularly updated. The digital data is made possible by technology (IT application), while the regular updating is facilitated by government-sanctioned citizen participation. Figure 4 shows how the data has undergone digitisation as vital steps in the implementation of the open data (the second process).

Figure 4. Steps in relation to data in data revolution (Second Process)

Data in non-digital form → Data in digital form → Data in digital form and updated

Digitisation   Data restructuring

---

10 The data revolution conducted in Bojonegoro district is an activity that consists of: (1) digitalisation of datasets on citizens and households (families) in all villages of Bojonegoro district; (2) providing additional information of geospatial location and photographs of every family by geo-tagging and using application on smartphone; and (3) providing access to citizen organisations at sub-village level to update individual data (citizen and family data) on their sub-district that represents actual change that occurs to individual citizens or families.
11 Bojonegoro district is selected as one of 14 sub-national governments participating in OGP pilot projects.

Table 1. Comparison between open data 1 and open data 2

	The first (early) open data initiative	The second (follow-up) open data initiative
Format	Some datasets were open and published in the national open data website (government website). By opening and sharing the data to the public, the government expected that the public would respond and use the data.	Data revolution: digitisation of single data, adding geo-tagging information and photos, collecting, storing, and processing data in the single (disaggregated) format. Open data contract: publication of contract data, the collection and processing of project's data in single data format.
Stakeholder involvement	The regent, the staff of the central government, the staff of the local government (especially the heads of the local agencies)	The regent, local government staff, village government, village cadres, NGOs
Which data is opened	All data available in government data centres, regardless of whether the data is updated or not. The data is mostly represented as aggregate.	The data which is collected with the participation of citizens in the form of single data and it is updated regularly based on citizen reporting.
Expected data utilisation	The government (or other agencies) and the citizens/ communities are expected to use the data as their references.	The government (or other agencies) and the citizens/communities are expected to use the data for further process into information that they need.

One of the important points in data revolution is related to the fact that data updating can be done immediately and the fact that the citizens can participate in data updating. Immediate and participatory updating are made possible because data revolution is about management and processing of data in a single (disaggregated) format. As explained above, the single data format allowed the updating and verification process to be applied to single individuals and households. On the other hand, the aggregated format, which is currently available in the local government data system, limits the management and updating process. Consequently, the utilisation of such aggregated data is also limited to being just a reference (without further processing according to the specific needs of the users).

Single data

To get a sense of what single data means, see representations of two data below, namely local poverty data (table of aggregated data) and family data (table of single data).

98

Table A: Local Poverty Data

Name of Village	Number of Poor Families	Number of Non-Poor Families	Percentage of Poor Families
Asri Jaya	271	544	33.25%
Bauran	67	288	18.87%
Cawanan	382	108	77.95%

Table B: Family data

Name of Family	Poverty Status
Amin	Poor
Bobby	Not Poor
Charles	Not Poor
Dody	Poor
......	......

Table A is the aggregation of data in Table B. To update data in Table A, which is in aggregated form, is difficult because it demands survey on all of the data. To illustrate the point: in Asri Jaya, the number of poor families is 271; however, we have no idea which individual families belong to the poor category. Only another survey can tell.

On the other hand, updating data in Table B is easier because it contains single data per family. Changing data on one family automatically modifies the aggregate data. As an illustration, the Amin family climb up the economic ladder (e.g. the head of the family has a better paying job), hence, the family stops being categorised as 'Poor' and automatically referred to as 'Not Poor' – modifying the aggregate data of poor families.

When data management operations are carried out on the table that contains single data representation (such as Table B), it is easier to update data and get a better understanding of the data, compared to updating on aggregated data (such as in Table A). With the use of a computer to conduct aggregation, updating is made easier. Changing single data automatically translates into modified aggregated data.

Observed change

Data revolution implementation produced some observed changes affecting the government and citizens in Bojonegoro district. It covers changes in the way they view and judge the quality of their data, changes in the way they use the data, and changes in the form of collaboration between the government and CSOs regarding data provision and data utilisation.

99

Some local government staff were aware of the data quality that they had and desired to improve the quality of their agencies' data and make it more open. For example, the local planning agency realised that their data on poverty was inadequate to address the real poverty challenges due to lack of detail about data on poor families (which ones are poor) and the slow update of the data. In contrast, data from the data revolution addresses those issues and is more suitable to their needs.

Local development actors (local government, village government, CSOs) started to have new insights about the utilisation of data. They now saw that the data is vital in measuring local development accomplishments as well as in improving targeted social assistance programmes in development. All of these have been possible due to the data revolution with its single (disaggregated) and open data.

There is a desire of the local government to collaborate with CSOs to improve data provision. The local government viewed the CSOs as capable in helping to provide disaggregated data. With this new government approach, institutions working to empower citizens (especially CSOs) are strengthened because of the opportunities presented by their involvement in data provision.

In our observation, it was the implementation of data revolution that allowed those changes to take place. We observed that the contribution of data revolution on those changes was as follows:

Figure 5. Contribution of the data revolution in stimulating data use

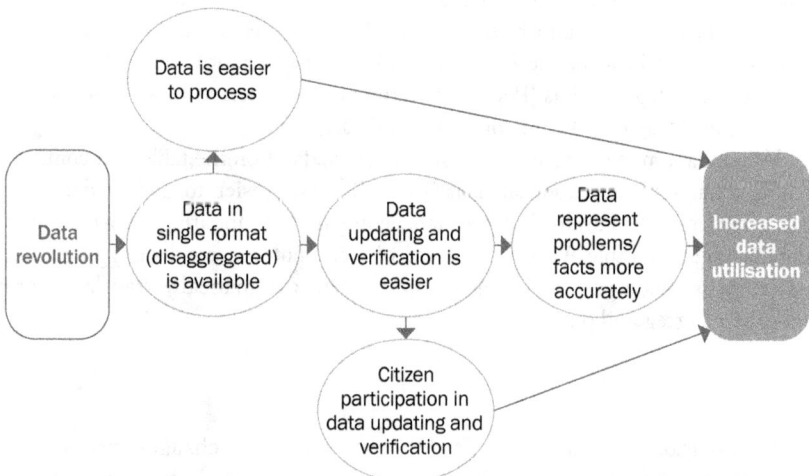

- Some local government staff of Bojonegoro welcomed the second process of open data implementation (data revolution) as they saw the possibility and feasibility of collecting and processing single data. Data in this

format can be used and processed into up to date information that the government needs. This inspired them to initiate a new treatment of their agencies' data with the same approach.

- Some lower level government officials (village level) found that the data revolution allowed gave them the opportunity to manage data easier, especially in data updating and verification. The single data made it possible to update and verify the data, which is something the officials needed.[12]

- The CSOs that we met also looked at the single data provided by data revolution as valuable. Owing to its updated nature, the data system allows for participation and makes transparency more useful. Some CSOs see that such data can be used to contribute to achieving development goals, for example, the data can be used for the monitoring of SDG accomplishments and for the monitoring of procurement contracts.

- The data revolution processes, in which there are components of significant participation by citizens/communities in the updating and verification of data, helped facilitate government-CSO collaboration. The experience of data revolution implementation resulted in a desire to maintain collaboration between government and CSOs in data-related work on a continual basis.

Potential for follow-up on the changes

The observed changes mentioned above have the potential to strengthen efforts toward reaching the intended impact of open data initiatives from the very beginning – data openness and utilisation of the opened data.

Figure 6. Expected scenario (based on changes already taking place)

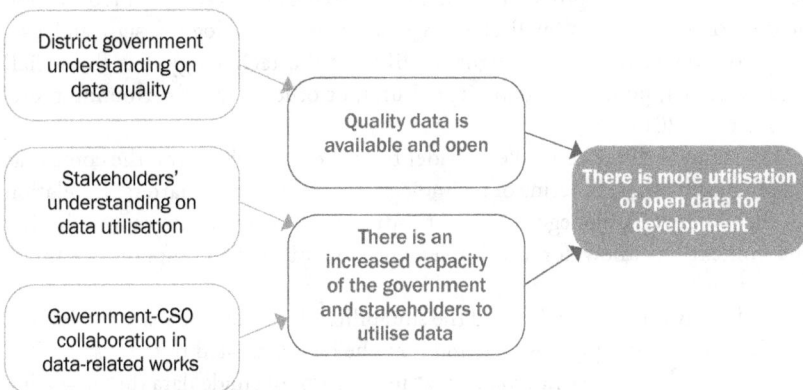

---

12   Village officials are responsible for targeted social assistance programmes funded by the upper level governments. The accurate, updated, and valid (verified) data are vital for them.

At the time of writing of this chapter, data revolution implementation is still ongoing, measuring impact/results at this point may be inapplicable or at least premature. What is presented as findings are the changes already taking place. Figure 6 above shows the expectations – a kind of future scenario – resulting from those changes. It is expected that the provision of open and good quality data becomes the agenda of the local government along with their increased understanding of the importance of data quality for development. In addition, the improved understanding of the stakeholders on the benefits of open data and the collaboration between government and CSOs in data-related works is expected to lead to or have implication in the form of the increased capacity of the government and other stakeholders in data utilisation.

Conclusion

The initial open data implementation in Bojonegoro district was carried out with a high level of transparency. Nevertheless, the expected level of data usage by the target groups remained low. This was inseparable from some disconnections that beset the implementation of the initiative. Our action research shows that the disconnections were caused by incompatibility of available data with the needs of the prospective data users; the quality of available data falling short of the quality required by the data users; and the lack of capacity of the data users in utilising the data.

This kind of phenomenon is common in ICT projects for development around the world, especially those in developing countries, where lack of data quality and low human capacity to initiate demand of governance reinforce each other. Diniz et al. (2014) noted that various studies locate the 'failures' of ICT projects at the gap between the need of the end-users and the perception of the ICT designers. They proposed an approach that emphasises the modification of the context, as it is 'malleable' in relation to technology design and use. They considered that 'users might modify not the technology, but the social, organisational, political, economic, cultural, or other context surrounding use' (Diniz et al. 2014: 16).

We stand to differ here. We consider both the technology and the context as shaping each other. In terms of technology, the data format matters. In relation to the context, Bojonegoro district introduced many changes (modifying the context) to allow for the implementation of ICT to support its Open Government Partnership goals.

That's why the data quality and the suitability of the data are inseparable from the format of data representation and how the data is treated in the system. The absence of a data representation format in the form of single data (because what is stored is only those in aggregate data format) led to data that is inflexible, difficult to update and hard to verify. Such a situation led the data users to consider and perceive the data as of low quality and less suitable for their needs.

The situation is considered as 'gaps' which must be closed with other initiatives or innovations. The initiative/innovation that the Bojonegoro district government took – referred to as data revolution – was an open data initiative that involved data-related works using single data and employing citizen participation in data collection. In this case, data revolution with its main feature of disaggregated data, use of IT applications, and the involvement of public participation in data-related works, were carried out to close the gaps.

Even though it is still too early to tell the impact of the implementation of the 'close the gap' solution, there have been promising changes in local government bureaucracy. For example, we found a new understanding and awareness from district government officials regarding data quality, the desire to utilise the opened data (as long as the data quality is improved), and there have been collaborations with CSOs in conducting data-related works.

In particular, we want to highlight the collaboration between CSOs and local governments in data-related works. The collaboration took place before and after the implementation of the data revolution initiative. CSOs in this case play a role in contributing to the conception of ideas and in implementation processes. At the time of the conception of the idea of the data revolution, CSOs took the role of conducting studies and formulating ideas for obtaining a data revolution model suitable for Bojonegoro district. While in the implementation phase, CSOs played a role in facilitating the participation of citizens in the data collection process.

The implementation of the data revolution initiative (or open data phase 2) is expected to have a direct contribution to the beneficiaries of local development – the citizens. But at this moment, it is still too early to say that the changes have reached that level. Nevertheless, there have been significant changes at the level of district government officials whose daily jobs are strongly related to local development data. This open data initiative allowed work processes using data to become easier, more democratic and more direct. Previously, if a government agency needed data from other agencies, then the process could take a long time, having to go through a long bureaucracy and with uncertainty about data quality. This open data initiative makes data from various agencies available on the web and can be downloaded immediately, although there are still quality problems as mentioned earlier. In addition, the introduction of a single data concept has made a difference in the way they perceive data quality.

In general, the lesson that can be drawn from the action research carried out by Sinergantara is that the availability of open data (in a government) does not always translate to the automatic benefit of open data access. To guarantee that open data does bring benefits, it is necessary to ensure data quality sufficiency (e.g. the data must be up to date and complete) and that the data can easily be processed to suit the needs of prospective data users. That is why it is crucial to take into account the format/representation of the data.

In particular, the lesson is related to data format/representation. The single data format introduced in phase 2 of open data in Bojonegoro (data revolution) showed a strong prospect to be followed up on. This single data format had raised the interest of development actors in the district, especially the government and CSOs.

REFERENCES

Diniz EH, Bailey DE & Sholler D (2014) Achieving ICT4D project success by altering context, not technology. *Information Technologies & International Development* 10(4): 15–29

Fox J (2016) Scaling accountability through vertically integrated civil society monitoring and advocacy. *Making All Voices Count Working Paper*. Brighton, UK: IDS

Gaventa J (2006) Finding the spaces for change: A power analysis. *IDS Bulletin* 37(6): 23–33

Kemmis S, McTaggart R & Nixon R (2014) *The Action Research Planner: Doing Critical Participatory Action Research*. London: Springer

Kindon S, Pain R & Kesby M (eds) (2007) *Participatory Action Research Approaches and Methods: Connecting People, Participation and Place*. Abingdon, Oxfordshire: Routledge

Maail AG (2017) The relational impact of open data intermediation: Experience from Indonesia and the Philippines. In: F van Schalkwyk, S Verhulst, G Magalhães, J Pane & J Walker (eds) *The Social Dynamics of Open Data*. Cape Town: African Minds. pp. 153–166

McNiff J & Whitehead J (2002). *Action research: Principles and practice*. London: Routledge

Open ePolicy Group (2005) Roadmap for open ICT ecosystems. https://cyber.harvard.edu/epolicy/roadmap.pdf

Orlikowski WJ (1992) The duality of technology: Rethinking the concept of technology in organizations. *Organization Science* 3(3): 398–427

Redman TC (2008) *Data Driven: Profiting from Your Most Important Business Asset*. Boston: Harvard Business Press

Roberts T & Hernandez K (2017) *The techno-centric gaze: Incorporating citizen participation technologies into participatory governance processes in the Philippines*. Making All Voices Count Research Report. Brighton: IDS

Sachs JD (2015a, 6 May) Data for development. Project Syndicate. https://www.project-syndicate.org/commentary/sustainable-development-data-by-jeffrey-d-sachs-2015-05

Sachs JD et al. (2015b, 18 September) The data revolution for sustainable development. Project Syndicate. https://www.project-syndicate.org/commentary/sdgs-data-collection by-jeffrey-d-sachs-et-al-2015-09

Smith M & Laurent E (2010) Open ICT ecosystems transforming the developing world. *Information Technologies and International Development* 6(1): 65–71

Srimarga IC, Wibowo A & Aisah S (2017) Efektifitas Pemanfaatan Teknologi Informasi Dan Komunikasi untuk Memperkuat Transparansi, Akuntabilitas dan Partisipasi serta Keselarasan Interaksinya dengan Platform Offline. Unpublished MAVC Learning Paper. Jakarta: Sinergantara

Yuliar S (2009) *Tata Kelola Teknologi: Perspektif Teori Jaringan Aktor*. Bandung: Bandung Institute of Technology (ITB) Press

6.

The cost of late payments in public procurement

Juan Pane, Camila Salazar & Julio Paciello

Open public procurement data has gained attention in the past years as a way of increasing transparency and helping governments improve their procurement practices. According to the latest report by the Open Contracting Partnership (2017: 3), in 2017 over 30 countries were publishing data following the open contracting data standard. Our analysis is a case study of how open data of public procurement, when analysed and re-used, can yield useful inputs to identify inefficiencies in the procurement process, recommend structural reforms to improve the procedures and drive cost savings for the public administration.

Following the Open Contracting Data Standard (OCDS), the publication of contracting data should include the different stages of the procurement process that go from the bid preparation (planning) and tender, to the implementation of the contract. For this chapter, we focus on the last stage, particularly the financial implementation which relates to the process of payment to suppliers. In this phase, governments sometimes face the problem of not paying suppliers on time, which causes inefficiencies and financial costs. Failure to pay invoices promptly generates a negative effect on firms that provide goods, services and public works to the government, causing short-term liquidity problems. This might force businesses to turn to the financial market to cover their obligations, use their savings, or go out of business (World Bank 2017). If money is lent from the financial market, providers incur extra costs due to interest rates, which might be included in the cost structure of the provider, implying an extra cost for the public administration. In addition, this practice can ultimately affect economic growth (Checcherita-Westphal et al. 2016). Moreover, if this procedure becomes the norm, suppliers may decline to do business with the government, reducing competition, and thus the possibility of obtaining better value for money for the purchasing entity (World Bank 2017).

Previous work on public procurement payments has focused on measuring arrears in national accounts (Diamond & Schiller 1987) and on how governments can prevent and manage payment delays (Flynn & Pessoa 2014). There are also specific country reports that analyse public procurement payment systems (World Bank 2008, 2017; Giussani et al. 2016). On the other hand, Checherita-Westphal et al. (2016) use annual data from 17 European countries and a proxy for government arrears to calculate the economic impact; and Connell (2014) estimates the cost of late payments on government to business transactions for 26 European countries. Other works have studied inefficiencies in other stages of the procurement process (Balaeva & Yakovlev 2017; Fiordelisi et al. 2012) or analysed late payments in the private sector (Smirnov 2016).

This chapter estimates the financial cost of late payments in public procurement and identifies variables that can affect them. Contrary to previous work that used proxy variables to calculate the cost, we analysed 599 354 detailed payments of 59 public institutions from 2011 to 2017 released using open data formats from the National Treasury of Paraguay (datos.hacienda.gov.py). We used a set of methodologies for cost analysis readily available in the literature and adapted them to the data availability, to show how this can be replicated in different countries, using the Paraguayan data as a case study. Our calculation follows Connell's (2014) approach, to estimate the cost of late payments on firms and on the public administration but focusing on a single country (Paraguay) for a seven-year period, given the rich information about the payment process released at datos.hacienda.gov.py.[1] In addition, we use survival analysis, a technique that allows us to model payment duration and to identify which variables have an impact in delaying payments. The preliminary findings show that the median duration from the moment an invoice is issued to when the payment occurs via a bank transfer to the supplier's bank account can be of approximately 55 days for each payment. In comparison, international practice considers 30 days an acceptable payment period (Flynn & Pessoa 2014). Moreover, our analysis on the historical data shows that the cost of late payments is of approximately USD 81.07 million in the analysed time frame, which could be cut down if some of the steps in the payment process are analysed and the appropriate normative framework is created.

Our work aims to encourage the publication of open public procurement payment data by showing how this information can be analysed to reduce costs in the public procurement process. Our methodology could be replicated in other

---

1   We used the following datasets: payroll of officials (https://datos.hacienda.gov.py/data/nomina), national budget (https://datos.hacienda.gov.py/data/pgn-gasto), list of invoices (https://datos.hacienda.gov.py/data/obligacion), transfer request orders (https://datos.hacienda.gov.py/data/str), transfer orders (https://datos.hacienda.gov.py/data/orden-transferencia).

countries that have already published payment data in open formats. Some of the limitations of our study are that we lack the exact invoice due date, specific information about the providers and that we do not have payment information about all the public institutions in the country.

This chapter is structured as follows: section 2 describes the state of the art related to our problem of study; section 3 explains the data and methods used for the analysis, transformations made to the variables and limitations of the study. Section 4 shows the results of the analysis and section 5 presents the conclusions and future work.

State of the art

The current state of the art on public procurement payments has mostly focused on discussing the issue of government arrears, on measuring the economic impact and financial cost of late payments for several countries and on analysing other stages of the procurement process.

The problem of arrears

The International Monetary Fund (IMF) has addressed the issue of late payments in different reports. Flynn and Pessoa (2014) analysed how governments can prevent and manage expenditure arrears, which includes payments to private contractors. According to the report, one of the biggest issues on managing arrears is that what is considered a delay may vary between countries and is dependent on the maturity of the payment system. They also list the impact chronic arrears may have at an aggregated level, which includes reduced economic growth, increased cost of service provision, reduced or interrupted public delivery and increased interest rates, among others. However, these effects are discussed theoretically and not measured empirically.

On the other hand, Ramos (1998) focused on how governments can address the arrears problem with securitisation in order to provide temporary relief from debt service obligations and increase government credibility.

For the specific case of Paraguay, the World Bank (2008: 62) pointed out in a fiduciary assessment that improvements were needed in the public procurement payment management system, since a relevant percentage of contracts were in arrears: 'The delay in effecting payment negatively impacts the willingness to participate in public PR processes or is reflected in quoting higher prices, discounting the financial costs associated to these delays, thus producing inefficiencies to the system'. In 2016, another assessment (Giussani et al.) pointed out that the volume of arrears (of all government expenditure) was around 9.68% of total expenditure, when the recommended percentage is of 2%. In addition, they cited a report by the public comptroller, which had estimated a 20-day average delay.

More recently, a World Bank's public procurement report (2017) presented comparable data on public procurement regulations across 180 economies; they divided the analysis in eight pillars, one of which was the payment of suppliers. In this category, Paraguay received a score of 48 out of 100 – the lowest score in all the indicators measured, since the actual time for suppliers to receive payment was between 31 and 90 days. In addition, there were no automatic penalties paid to the suppliers in the case of a delay and the time to process the payment did not start from the submission of invoice, which can slow down the payment arbitrarily.

Cost and impact of late payments

To the best of our knowledge, the cost of late payments in the public sector has yet to be explored in Latin America, and recent studies have measured the economic impact of late payments in European countries. For instance, Checcherita-Westphal et al. (2016) used annual data from 17 European countries and a proxy for government arrears to calculate its economic impact. They discovered that payment delays reduce economic growth, increase the likelihood of bankruptcies and reduce profits. For instance, a one standard deviation change in delayed payments reduces the growth rate by 0.8 to 1.5 percentage points and reduces profit growth by 1.5 to 3.4 percentage points. Additionally, they found that the larger the delayed payments, the higher the probability of default among private companies. Because of the cross-country and short time period data, they used dynamic panel models and a Bayesian VAR to estimate these effects.

Additionally, Connell (2014) estimated the cost of late payments on government to business transactions at an aggregated level, also in the European context. The study calculated the short-term financial cost, applying annual interest rates to the claims against the public administration (which serves as a proxy for the payment delays), times the average delay. This approach to calculate aggregated costs is useful for our estimation, with the difference being that we focus on a single country. Furthermore, Connell (2014) calculated the impact these delays may have on the exit rate of firms, using the ratio between the number of deaths of enterprises and the total number of firms in a given country, across several years. The result shows that a 1-point reduction in the delay ratio leads to a decrease in exit rates of about 1.7 percentage points.

A limitation found in these studies is the lack of full information regarding government payments, so the estimation is made using a proxy. On the contrary, we have micro-level data about each payment which allows us to calculate the aggregate cost more precisely.

Furthermore, Fiordelisi et al. (2012) calculated the cost for the Italian economic system resulting in the delay of trade loans by the public administration. The study explored different scenarios according to different payment times and estimated the financial cost using the interest rate multiplied by the delay and volume of credits towards the public administration. In addition, they estimated the aggregate social cost comparing this value with the expenditure the government would have faced to pay on time, using the rate on Treasury bills.

On the other hand, Valcani Vicari Associati et al. (2015) evaluated whether the 2011 European Union Late Payment Directive had accomplished its objective of reducing payment times in 28 countries in Europe. They found that after four years of implementation, the average duration had fallen from 65 days on average in 2011 to 58 days in 2014, but stayed beyond the 30-day optimal deadline: 'Rather than legislation, national business culture, economic conditions and power imbalances are the driving factors for payment behaviour' (2015: 68). This is a relevant result for our investigation, since it is important to consider other factors besides legislation in order to drive change in the payment culture. They also found that firms that had the government as a main client were more likely to have difficulties paying to their suppliers, which indicates that late public payments can have an impact on a larger supply chain.

Inefficiencies in public procurement

The literature on public procurement also includes several studies which analyse the cost and efficiency of the public procurement process, and not exclusively of the payment stage. Some of these works use data envelopment analysis (DEA), a method that is generally used to estimate a production cost function, with minimal assumptions. This methodology calculates an efficiency frontier for a set of units using different inputs and outputs, and then gives an efficiency score (it considers efficiency as the lowest input amount to produce one unit of output). Guccio et al. (2012) used this method to investigate the performance of Italian public contracts in terms of time of completion of works and cost overruns.

We find more useful the approach used by Strand et al. (2011) and Balaeva and Yakovlev (2017). They analyse the cost and effectiveness of public procurement in the European Union and Russia using estimates of person-days spent in the procurement process, and then they applied data of employee remuneration to calculate the labour costs and the total costs of the procurement. In their regard, shorter procedure times indicate higher efficiency. This methodology can be adapted to analyse exclusively the payment stage of the process. However, we could not implement this methodology due to lack of data.

Finally, another set of studies have analysed delays in other stages of the public procurement process. For instance, Gori et al. (2017) explored the variables that affect the duration of public works in Italy using survival analysis and found that the lack of experience of local governments results in a higher delay probability in the execution of the contract. On the other hand, Smirnov (2016) proved survival analysis is a good approach to model late invoice payment times in the private sector. Given the similarity of the transaction, this approach can be used to model government to business payments.

Methodology

Data description

The public procurement payment process in Paraguay consists of three different stages. The first one comprises the period from when the invoice is issued until it is approved by the procuring entity (Invoice stage). In the second stage, a transfer request (TR stage) is generated, and it is then sent to the Treasury. In the final stage, the Treasury creates a transfer order (TO stage) and then makes the payment to the contractor via bank transfer. The first two stages happen inside the procuring entity, while the last stage is the charge of the Treasury. There are also intermediate stages in each of the steps, illustrated in Figure 1.

Figure 1. Payment process steps

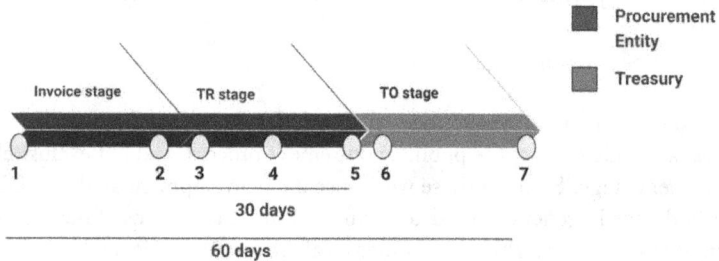

Where:

1. Invoice creation (receipt of invoice) *(obl_fecha elaboracion)*[2]
2. Invoice approval *(obl_fecha_aprobacion)*
3. TR creation *(str_fecha_ingreso)*
4. TR approval *(str_fecha_aprobacion)*
5. TR transfer to Treasury *(str_fecha_recepcion_tesoro)*
6. TO creation *(ot_fecha_generacion)*
7. Payment *(ot_fecha_deposito)*

The process is as follows: the provider presents the invoice to the procurement entity (1) and then has to wait until the institution creates an obligation for the invoice and approves it (2). For one or more obligations, a transfer request (TR) is created (3), it is then approved (4) and finally the procurement entity sends it to the Ministry of Finance (Treasury) for payment (5). Therefore, for one or more TR's the Treasury creates a transfer order (TO) (6), and then executes the payment (7).

---

2   In parenthesis, we add the variable name derived from the dataset. All the variables are explained in Table 1.

According to the IMF, 'international practice on what is an acceptable delay between receipt and payment of the invoices varies from anywhere between 30 to 120 days' (Flynn & Pessoa 2014: 4). In addition, the World Bank (2017) reported that most suppliers in high-income economies receive payments in less than 30 days.

In Paraguay, there is not a single deadline in laws or regulations, from receipt of invoice to payment, and the specific deadline is stipulated in each contract. However, according to the executive order 8452/2018, the TR stage has a 30-day deadline. In addition, the DNCP order 1024/11 about General Conditions of the Contract (Pliego estándar de contratación) states that the contracting party will execute the payments as soon as possible, but in no case may it exceed sixty days after receipt of invoice or request of payment, and after the contracting party has accepted the request. The acceptance or rejection of the invoice must be given no later than fifteen days after its presentation.[3] These 15 days are before the process starts and the invoice is created. This means that the whole payment process can last 60 days, if not specified otherwise in the contract. In addition, according to Law 2051 and the General Conditions of the Contract, if the payment is delayed, the contractor has to recognise financial interests to the provider, which implies an additional cost for the public administration.

According to this legislation, the regulation of payment times is not clear, which can contribute to long costly delays. For our analysis, we consider the following delays and deadlines:

Table 1. Stages and deadlines of the payment process

Stage/Delay	Description	Deadline in regulations
Invoice stage	Time between the issue of the invoice and its approval in the procuring entity. (2) – (1)[4]	No specific deadline
TR stage	Time between the issue of the TR and its dispatch to the Treasury. (5) – (3)	30 days
TR creation delay	Time between the approval of the obligation (invoice) in the procuring entity and the TR creation. (3) – (2)	No specific deadline
TR approval delay	Time between the TR creation and its approval. (4) – (3)	No specific deadline
Transfer to Treasury delay	Time between the TR's approval and its dispatch to the Treasury. (5) – (4)	No specific deadline
TO stage	Time between the issue of the transfer order and the final payment. (7) – (6)	No specific deadline
TO creation delay	Time between the TR's dispatch to the Treasury and the creation of the TO. (6) – (5)	No specific deadline
Payment delay	Time between the creation of the TO and the final payment. (7) – (6)	No specific deadline
Total delay	Time between the issue of the invoice and the final payment. (7) – (1)	60 days

3   Own translation.
4   Steps detailed in Figure 1.

For this analysis, we used the following datasets of public institution payments between 2011 to 2017 in Paraguay, available in the open data portal of the treasury (datos.hacienda.gov.py):

- List of obligations (Listado de obligaciones, available at https://datos. hacienda.gov.py/data/obligacion), a dataset with information about each invoice;
- Transfer request (Solicitud de transferencia de recursos, available at https://datos.hacienda.gov.py/data/str), a dataset with all the requests that are generated in the entity for one or several payment obligations to the provider. Each transfer request (TR) can have one or more invoices; and
- List of transfer orders (Listado de órdenes de transferencia, available at https://datos.hacienda.gov.py/data/orden-transferencia), which has the detail of each transfer order, an instrument through which the delivery of funds to the beneficiary is made. Each transfer order can have one or more transfer requests.

The above datasets were combined using a unique identifier, where the observation unit was given by each invoice. The dataset had information about each invoice, the contracting procedure and the provider. We also used two other datasets from the treasury's open data portal to obtain two variables about the institutional budget execution[5] and the number of officials working in each entity.[6] These served as proxies for the institution size and its efficiency using public resources.

The initial dataset had 599 369 observations. However, we found several errors, which we validated with the Treasury officials. In the cases where these errors could not be corrected, we eliminated the observations. We made the following transformations to clean the data:[7]

- We eliminated the observations with empty date values and those in which the date of the invoice was before the date of the contract.
- After creating variables for the delays, we obtained several negative values in the duration. We eliminated these observations, since we could not correct the error.
- We eliminated observations where the duration of the payment was less than 10 days, an unlikely scenario; and observations where the sum of the invoice was less than 50 000 guaranies (USD 9).

---

5   https://datos.hacienda.gov.py/data/pgn-gasto
6   https://datos.hacienda.gov.py/data/nomina
7   See section 3 in http://rpubs.com/camilamila/late_payments

In addition, we identified outliers[8] in each of the stages of the process and created two other datasets:

- Dataset A, the main dataset without the outliers and 84% (315 983) of the observations.[9]
- Dataset B, a dataset with the outliers and 61 537 observations. In this dataset, 55% of the observations were from the Ministry of Health.

We decided to segment the dataset into two, since the longer delays we have in dataset B, correspond to less frequent cases, that do not reflect the payment practices of most of the institutions in the sample. However, all the analyses were run on the full dataset and in some cases, we present the results separating by the dataset; whenever this occurred, we specified it in each of the tables and the figures in the Results section.

After the transformation, the datasets had the following variables:

Table 2. Variables used in the analysis

Variable name	Description	Type
obl_id	Unique id for each invoice	string
obl_fecha elaboracion	Invoice creation date	POSIXct
obl_fecha_aprobacion	Invoice approval date	POSIXct
str_fecha_ingreso	TR creation date	POSIXct
str_fecha_aprobacion	TR approval date	POSIXct
str_fecha_recepcion_tesoro	TR transfer to Treasury date	POSIXct
ot_fecha_generacion	TO creation date	POSIXct
ot_fecha_deposito	Payment date	POSIXct
obl_monto_obligado	Amount of each invoice (includes deductions)	num
obl_anho_obligacion	Year of the invoice	num
obl_codigo_contratacion	Code of the contracting procedure	string
obl_moneda_descripcion	Invoice currency	string
obl_entidad_descripcion	Procuring entity	string
funds	Type of funds used to pay the invoice: 1=Institutional funds, 2=Treasury funds, 3=public credit funds	factor
purchase	Type of purchase: 1=services, 2=goods and materials, 3=exchange goods, 4=investment, 5=transfers	factor
contr_monto_adjudicado	Total amount of the contract to which the invoice belongs	num
contr_fecha_firma_contrato	Date of signature of the contract	POSIXct
prov_ruc	Unique id of the provider	string
total	Total duration of the payment in days	num
r_factura	Total duration of the invoice stage	num

---

8   A value was considered the outlier if it was higher than Q3 + 1.5 *IQR, where Q3 equals the 75th percentile of the distribution and IQR is the interquartile range.

9   The datasets used for the analysis can be downloaded from: https://drive.google.com/open?id=1Qhr_LPIoTyJa8Ll9kI_OnL-tIaoikjYn

Variable name	Description	Type
r_str	Total duration of the TR stage	num
r_ot	Total duration of the TO stage	num
r_carga_str	str_fecha_ingreso - obl_fecha_aprobacion	num
r_apro_str	str_fecha_aprobacion -str_fecha_ingreso	num
r_tesoro_str	str_fecha_recepcion_tesoro-str_fecha_aprobacion	num
r_carga_ot	ot_fecha_generacion-str_fecha_recepcion_tesoro	num
r_pago_ot	ot_fecha_deposito -ot_fecha_generacion	num
total_cat	categorical variable of payment times	factor
interes	average annual lending rate	num
deflator	PIB deflator, base 2017=100	num
official	Number of officials working in the procuring entity	num
ejecucion	Average annual budget execution percentage (2011-2017) of the procuring entity	num

Survival analysis

We modelled late invoice payments using survival analysis to identify which variables have a role in delaying payments. Survival analysis is a statistical method used to analyse and model the data when the outcome variable is the time until the occurrence of a specific event. In our case, the event of interest is the date of payment of an invoice. These methods have been previously used by researchers to study the duration of public works in Italy (Gori et al. 2017) and to model late invoice payments times in the private sector (Smirnov 2016).

The idea of this approach is that subjects are followed during a time period until the event occurs. The event is called a *failure*. In this case, t=0 is the date the invoice is created and the *failure* will occur when the payment happens. In the context of survival analysis *survival* will mean the invoice hasn't been paid yet.

The duration of the state (payment period) is a non-negative random variable called *T*, with a cumulative distribution F(t) (Cameron & Trivedi 2005). The probability that the duration of the episode is less than *t* is:

$$F(t) = \int_0^t f(s)ds = Prob(T \leq t) \quad (1)$$

The probability that the event equals or exceeds *t*, or in our case the probability that the invoice is not paid before time *t*, is given by the survival function:

$$S(t) = 1 - F(t) = Prob(T > t) \quad (2)$$

We also estimated the hazard function *h(t)*, which is the probability of leaving a state conditional on survival time *t*. That is to say, the rate of success at time *T=t* given that the invoice has not been paid for up to time *t*.

$$h(t) = \frac{Prob(t \le T \le t + \frac{\Delta t}{T} \ge t)}{\Delta t} = \frac{F(t + \Delta t) - f(t)}{\Delta t \, S(t)} = \frac{f(t)}{S(t)} \quad (3)$$

We used the Kaplan-Meier method to estimate the survival function, project the survival curves and estimate the differences between groups. The estimator $S(t)$ is given by:

$$S(t) = \prod_{t_i < t} \frac{n_i - d_i}{n_i} \quad (4)$$

where n_i is the number of survivors at time t_i and d_i is the number of events that happened until time t_i.

Finally, to estimate the effect of covariates in survival time, we calculated a parametric model with a Weibull distribution, after testing the fit of different distributions on the data.[10] We included the following variables as predictors:

- *sum* of each invoice (log transform);
- *contract sum*, (log transform);
- *institution*, in this case this will be the debtor;
- *number of officials*, size of the institution payroll;
- *budget execution*, average budget execution of the entity (contractor);
- *type of funds*, the financial source used to pay the invoice (categorical variable); and
- *type of purchase*, segmented in services, goods and materials, investment and others (categorical variable).

These analyses were run excluding the outliers (dataset A).

Financial aggregated cost estimation

To estimate the short-term aggregated financial cost of late payments for firms, we followed Connell's (2014) approach to calculating this effect on government-to-business transactions in Europe.

The idea behind this calculation is that private providers need to compensate for the lack of liquidity generated by payment delays. Our assumption is that failure to pay invoices promptly generates a negative effect on providers, causing short-term liquidity problems for firms, and forcing them to turn to the financial market to cover their obligations. In addition, this can be seen as an opportunity cost for firms since, even if they have enough liquidity to face delays, they cannot invest and have to use the money to face the costs of the delay. Moreover, national legislation indicates that providers are entitled to receive interest due to

---

10   See section 5 in: http://rpubs.com/camilamila/late_payments

delays, so we assume that if the payment was delayed the public administration had to pay this extra cost.

To obtain the financial cost, we used the annual average lending rate, which is the bank rate that meets the short and medium-term financing needs of the private sector, to the amount of the overdue payment, times the delay expressed as a fraction of a year:

$C = P \cdot i \cdot d$ (5)

where:

C= estimated cost for firms

P=amount of each payment in guaranies

i= annual average lending rate

d= delay expressed as a fraction of a year

We consider that a payment is delayed if it is paid after 30 days of the submission of the invoice, which is the maximum payment time that institutions should have according to international recommendations. In addition to the 30-day limit, we also consider the following thresholds for what is considered a delay, according to the deadlines specified in Table 1 and international recommendations:

- 45 days (proposed intermediate deadline);
- 60 days (current maximum deadline in local legislation); and
- 15 days (proposed deadline for the invoice stage).

We calculated the number of invoices that surpassed the mentioned deadlines and then estimated the cost of this delay. Moreover, while we consider that delays occur after 30 days, we assume that even payments that are executed before 30 days imply a cost for firms, since they have to compensate for the lack of liquidity during this period. Thus, we also calculated the total payment duration cost that considers not only late payments (more than 30 days) but the total payment duration. Finally, we also took into account different scenarios, with different payment times, to analyse the possible savings the public administration could have if it modifies its payment regulations.

The cost was then adjusted by inflation using the GDP deflator. Both the data of the average lending rate and the GDP deflator were obtained from the International Financial Statistics website.[11] The amounts were converted to USD using the average 2017 exchange rate (5 618.933) from the World Bank open data portal.[12] This approach can be applied to other countries, using these same sources of financial information and local procurement data.

---

11 Dataset available at: http://data.imf.org/?sk=4C514D48-B6BA-49ED-8AB9-52B0C1A0179B&sId=1409151240976

12 https://data.worldbank.org/indicator/PA.NUS.FCRF?locations=PY

Limitations

Lack of exact invoice due date

We did not have the exact due date for each of the invoices, since this data are specified in the contracts. This impedes the ability to calculate the exact due date on the invoices and thus the potential cost that the administration had to recognise in interests to providers. A recommendation to public authorities would be to include this information as part of the public procurement open data.

Information about the providers

Not having information about the characteristics of the firm (size, sector) can result in an underestimation of the cost, since the interest rates can be different. For simplicity, we used the average lending rate for the private sector. Moreover, having open data about beneficial ownership could help expand the analysis to determine if there are clusters of providers that accumulate contracts.

Institutional data availability

Even though we have a robust dataset with payment information of 59 institutions that pay through the treasury payment system, this is not a complete sample of all the public institutions, thus the cost is underestimated. The study could be expanded when more institutions provide payment information in open data formats. Finally, we had to discard many observations due to errors on the original dataset. This shows there is still an opportunity to improve the collection and transformation of procurement data in open data formats and implement other validation techniques in order to minimise the errors.

Results

Descriptive and survival analysis

Payment duration by stages and funds

From Table 3 it is determined that only 18.6% of the invoices issued for public contracts between 2011 and 2017 in Paraguay were paid in 30 days or less. The 30-day deadline is considered an ideal threshold in which payments to providers should be executed. This means that 81.4% of payments in the analysed period were delayed. On average, public institutions took 55 days to pay invoices.

117

Table 3. Invoices paid by payment times (full dataset)

Payment period (days)	n	Percentage
< 30	70 402	18.6%
30 – <45	67 554	17.9%
45 – <60	57 841	15.3%
60 – <75	44 468	11.8%
75 or more	137 255	36.4%

However, considering the 60-day deadline that payments can have according to regulations, if not specified otherwise on the contract, 48.2% of payments were delayed. This shows, that even though the payment deadline in Paraguay is twice the period of what is considered an optimal payment time, almost half of all public procurement payments in the analysed period surpassed that limit (see Figure 2). This practice affects providers and the public administration, since this cost might be internalised in the provider's cost structure, and thus the final price of the good, service or public work is higher than expected.

Figure 2. Payment duration

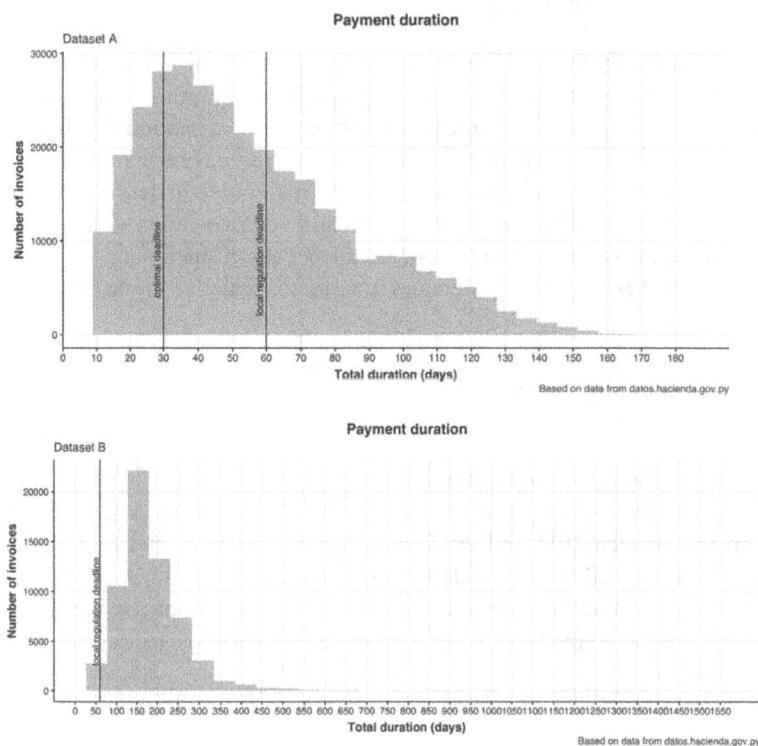

When segmented by year, we found that the median duration spiked after 2012 (see Table 4), it increased from 37 days to more than 52 for the subsequent years. Between 2011 and 2012, there were pay rises approved for public servants, which increased the spending in wages by 44% between 2011 and 2017. In contrast, the spending in the budget items related to public contracts only grew 11% on the analysed period. This caused liquidity problems for the public administration, which had more trouble assigning funds to pay contracts.

Table 4. Median payment duration by year (Dataset A)

Year	n	Median payment duration (days)
2011	33 894	34
2012	50 309	37
2013	31 970	58
2014	44 632	57
2015	50 868	56
2016	49 941	56
2017	54 369	52

This change is well illustrated in Figure 3, where we see an increase in payment times in the first and last stages of the payment process. While the invoice stage has the longest median duration of 22 days, the TO stage increased from a median of five days in 2011 to between 13 to 20 days in the subsequent years. In this last stage, the treasury has to execute the payments according to the available resources, which were more limited after 2012.

Figure 3. Payment duration by stage

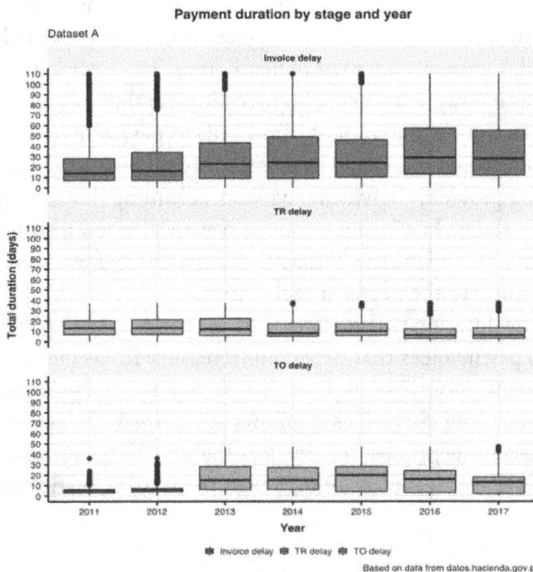

119

Figure 4 shows a more detailed picture of the change in average duration times by stage. For instance, it is clear that after 2012 procuring entities increased the time they spent loading the invoices in the system and approving the obligations, while the Treasury took longer creating a transfer order after they received a transfer request from the procuring entity.

Figure 4. Average duration by stage and year

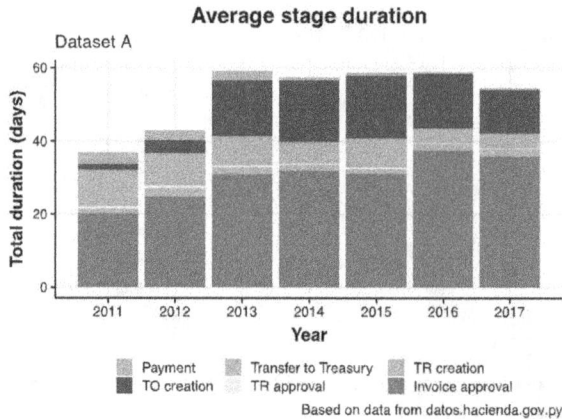

Moreover, the first stage of the process is the one that takes longer, regardless of the institution (see Figure 5). According to Treasury officials the main cause for this delay is that invoices cannot be loaded into the system and then approved if there is not a cash plan assigned by the Treasury, to pay those obligations. The main problem at this stage is that in practice there is a difference between the projected spending budget and the actual income budget that is financed through tax revenues, thus, the entities might not receive all their projected resources for a certain period. However, since the obligations are not recorded into the system, the Treasury has no way of knowing how to prioritise the assignment of resources to institutions that have a large number of bills to pay, which contributes to the delay. Our recommendation would be to improve cash planning and management, to reduce timing problems between payments becoming due and the availability of funds to pay them. For instance, if the system allowed loading the invoices prior to having a cash plan, the Treasury could identify how much resources to distribute according to the needs of each entity.

Moreover, the duration is also affected by the funds used to pay the invoices. It takes longer to pay invoices that depend on Treasury funds (55 days), than those using institutional resources (36 days). To explore this relationship further, we used the Kaplan-Meier curves. As it follows in Figure 6, the curves do indicate a difference between the groups: the invoices that use Treasury funds, take longer to be paid (higher survival probability), than those using institutional funds.

Figure 5. Average payment duration by institution

Average payment duration by stage
Top 10 entities with higher spending (full dataset)

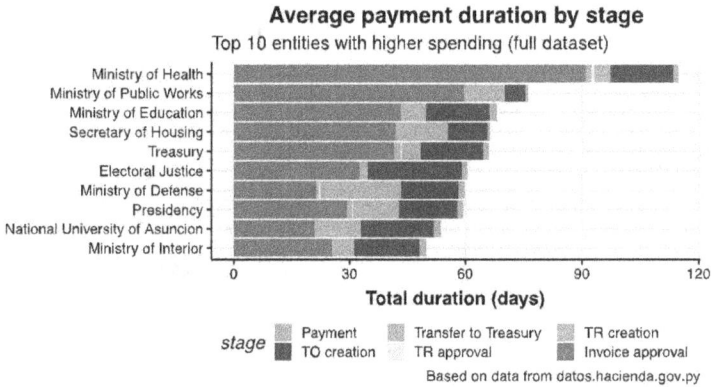

Based on data from datos.hacienda.gov.py

Figure 6. Payment duration by type of funds (Kaplan-Meier curves)

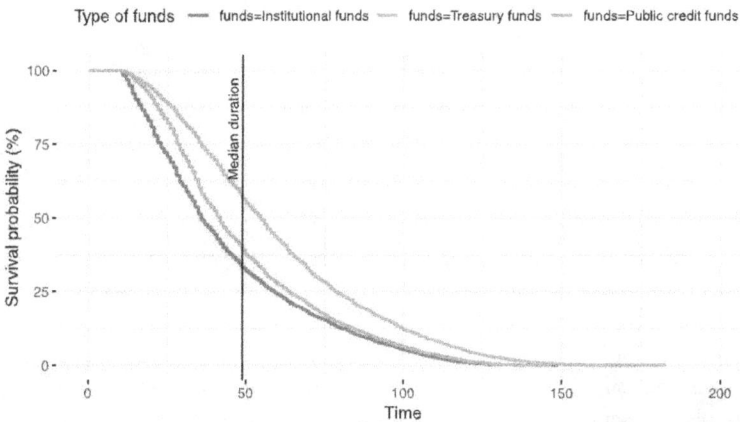

As shown in Figure 7, using Treasury funds delays two stages of the process: the invoice stage, since institutions cannot load the invoices into the system without a cash plan (as explained above), and the TO stage, since the Treasury cannot pay for the obligations if there are no available resources. On the contrary, bills that use public credit funds and institutional resources, do not need a cash plan to create obligations in the first stage, and once the transfer order reaches the Treasury, the funds are available to execute the payment. However, for these two types of funding, the invoice stage has a median duration of more than 22 days, which could be a sign of inefficiencies inside the procurement agencies and not a result of budget constraints.

Figure 7. Payment duration by type of funds

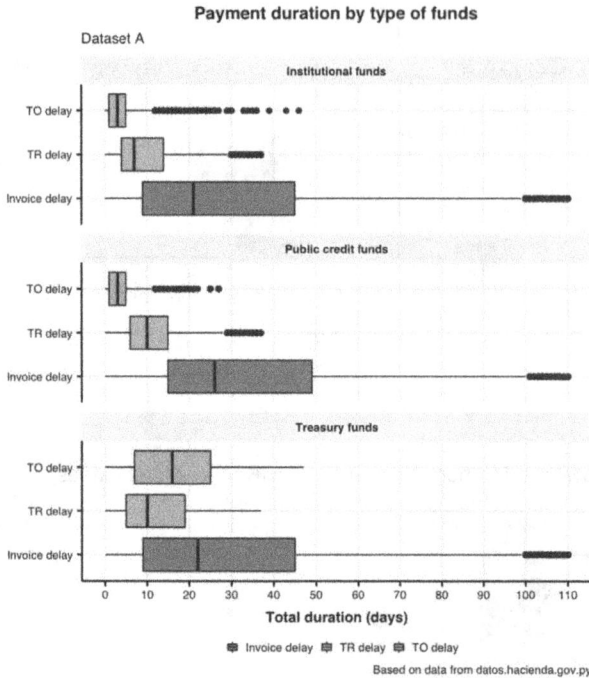

Payment duration by type of funds

Duration by institution

We divided the institutions by their size, according to their current payroll, to explore if the size of the entity could indicate a difference in payment times. Our hypothesis is that larger institutions (with more than 10 000 employees) have a larger volume of invoices to pay and might have more complicated bureaucratic processes, than smaller entities (less than 1 000 employees). As shown in Figure 8, the Kaplan-Meier curves show a significant difference between the groups. Large institutions have a higher survival probability (meaning they take longer to pay invoices) than smaller entities. While 50% of the invoices in small and median institutions are paid in less than 50 days, larger entities take longer.

When calculating the duration by institutions (in general), there seems to be important differences of payment times, which can be a sign of different practices between the entities. The Ministry of Health (the second in size) has a median payment duration of 73 days (without outliers), while the Foreign Ministry (median size) pays invoices with a median of 22 days. Only 6 of the 41 institutions analysed[13] have a median duration of less than 30 days.

13 For this part, we only analysed the institutions where data was available for the complete period 2011–2017.

122

Figure 8. Payment duration by institution size (Kaplan-Meier curves)

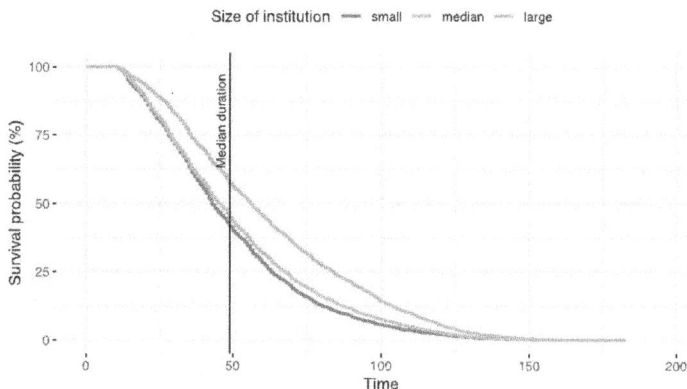

Table 5. Median and maximum duration by institution (top ten with higher spending)

Institution	Median duration (Dataset A)	Maximum duration (Dataset A)	Median duration (Dataset B)	Maximum duration (Dataset B)
Ministry of Health	73	183	191	698
Ministry of Finance	56	166	155	468
Ministry of Public Works	48	175	177	678
Presidency	48	175	144	508
Ministry of National Defence	47	180	138	753
National Secretary of Housing	47	160	184	492
Ministry of Education	45	155	154	408
National University of Asunción	40	169	122	485
Electoral Justice	35	154	104	419
Ministry of Interior	34	160	104	749

Moreover, besides having the longest median duration, the Ministry of Health is the second with the highest spending in payments, with an accumulated real spending of approximately USD 772 million (18% of the total), in the analysed period. Nevertheless, the delays in the Ministry of Public Works can be the costliest since this institution accounts for 45% of the total spending in public procurement payments. The difference in the duration of these two institutions could be the funds used to pay the invoices: while 68% of the Ministry of Health's invoices were paid using Treasury funds, the Ministry of Public Works paid 60.7% of its bills with funds from the Public Credit. As explained above, using Treasury funds causes delays in two stages of the process, however, Figures 9 and 10 show that both entities are inefficient at loading the invoices regardless of the type of funds, which contributes to extending the duration. Thus, it is necessary to evaluate the payment practices in these entities to find the cause for the delays.

Figure 9. Payment duration by type of funds

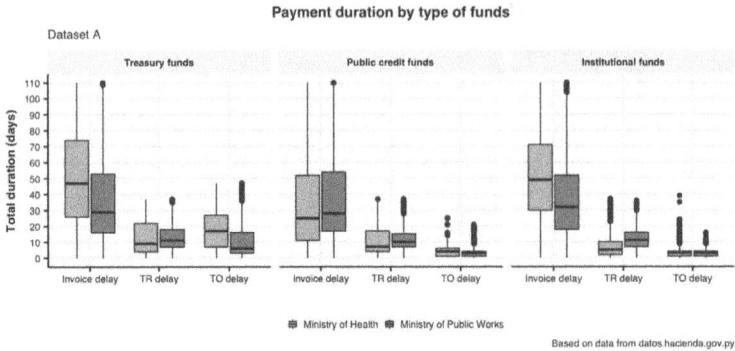

Payment duration by type of funds

Dataset A

Based on data from datos.hacienda.gov.py

Figure 10. Payment duration by funds and stages

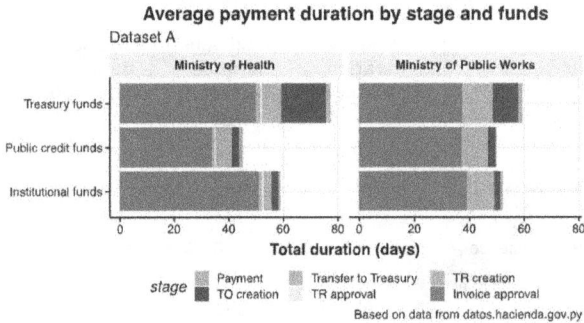

Average payment duration by stage and funds

Dataset A

Based on data from datos.hacienda.gov.py

Providers and type of purchase

On the other hand, out of the 4 742 providers that received payments in our sample, 80% of them had more than half their invoices paid after 30 days. Only 8% of the contractors were paid without a delay. Thus, payment delays are common and affect most of the participants in the procurement processes. In addition, for the larger sample, providers that receive transfers from the government suffer from the longest median duration of 70 days (see Table 6). This is a special case of the Ministry of Education that uses transfers to buy food for schools. In addition, goods and materials have a median duration of 56 days, while invoices related to services or infrastructure are paid faster, with a median of 47 and 43 days, respectively. For the extreme values dataset, the median duration of goods and services, of 177 days is similar to the one of investment works of 176. We would expect that smaller invoices and simpler contracts, like the ones related to goods and materials are easier to pay than works of infrastructure that tend to be costlier, however, this does not seem to be the norm in Paraguay's public procurement. This can be due to the funds

124

used to pay the invoices (as explained above) or that some providers might have more negotiation power when executing contracts and demanding payments. Nonetheless, the lack of information about the companies regarding their size, sector and financial characteristics, impedes a more detailed analysis and this could be explored in future works.

Table 6. Duration by type of purchase (Dataset A)

Type of purchase	n	Median duration	Maximum duration
Services	156,591	47	175
Goods and materials	128,209	56	180
Exchange goods	251	41	133
Investment	30,498	43	183
Transfers	434	70	155

Seasonal trends

We also observed seasonal trends and found that payments tend to increase in the last quarter of the year, before the end of the fiscal year in December (see Figure 11). In 2013, 2014 and 2015, there is also a peak in May that might be caused by the time it takes to execute payments. Thus, the invoices issued at the beginning of the year start to be paid at the end of the first quarter. In addition, there were no payments registered in January, in any of the years analysed, which can be due to lack of budget or liquidity in public institutions. In comparison, the issue of invoices peaks for most of the years between July and September. In general, there is a lag between the issue of invoices and payments, which confirms there are delays. Providers might be aware of this problem, and thus issue their invoices at particular times in order to guarantee payments before the end of the year.

Figure 11. Invoices billed and paid per month

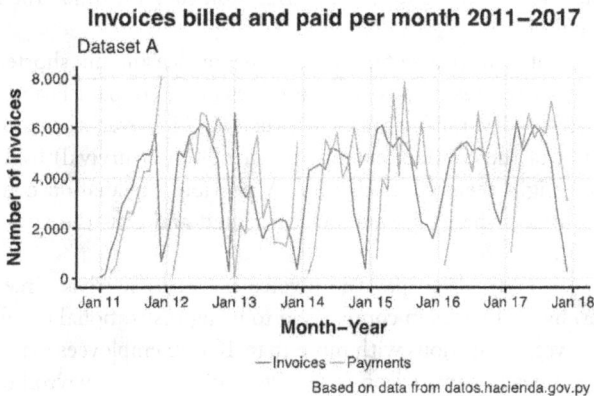

125

We also observed that invoices are accumulated before the Treasury pays them. As shown in Figure 12, the stock of unpaid bills is higher in the second and third quarters and decreases at the end of the year. This means that providers that present invoices at the beginning of the year face longer delays and thus higher costs than those with payments in the last months. In fact, the median duration of payments for bills issued in the first quarter is 66 days, while it reduces to 34 days for invoices presented after October; in the case of invoices presented in the second and third quarters, the median duration is 49 days. By taking into account the seasonal trends institutions can better plan their procurement processes, in order to schedule payments in periods where there is enough liquidity to pay promptly; or, if possible, distribute the stock of invoices through the year, instead of concentrating them in a particular period.

Figure 12. Unpaid invoices by year

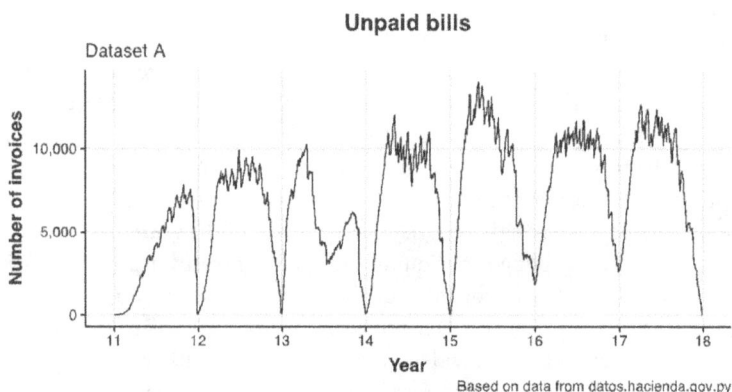

Based on data from datos.hacienda.gov.py

Survival model

Finally, we estimated a parametric model with a Weibull distribution, to calculate the effect of the different covariates on survival time. The results are presented in Table 6.

We found that a one per cent increase in the invoice amount shortens survival time by 0.97 times, this means that larger invoices are paid faster, under the assumption that all variables are held constant. However, the same increase in the amount of the contract extends payment time (survival) by 1.03 times, so the effect might seem contradictory. A possible explanation could be that for bigger contracts the payments are segmented and thus the invoices have a smaller amount.

Using Treasury funds to pay the invoice increases survival time (payment takes longer) by 1.31 times in comparison to using institutional or public credit funds. Moreover, institutions with more than 10 000 employees extend survival time by 1.43 times in comparison to smaller entities with a payroll of less than

1 000. As explained in our descriptive results, larger institutions have a higher volume of invoices and the payment process can be more complex. Also, an increase in the institutional budget execution reduces survival time, showing that entities that are more efficient in executing public resources tend to pay providers faster.

Table 6. Coefficient estimates

	Weibull model estimation		
	Coeff.	**exp(coeff.)**	
Invoice amount (log)	-0.0299	0.9705	***
Treasury funds	0.2747	1.3161	***
Institution size (base=small)			
Median	0.1983	1.2193	***
Large	0.3578	1.4302	***
Institutional budget execution	-1.5322	0.2161	***
Contract amount (log)	0.0311	1.0311	***
Type of purchase (base= Services)			
Goods and materials	0.0454	1.0465	***
Investment	0.0360	1.0367	***
Other	0.1513	1.1633	***
N	296299		
AIC	2771433		
Log-likelihood	-1385706		
Statistical significance	*** p<0.01		

Finally, invoices related to services purchased are paid faster than those related to investment or goods and materials. For instance, investment contracts extend payment time by 0.03 times, in comparison to services. Other previous work (Gori et al. 2017) had found that infrastructure contracts are associated with longer delays, and thus this could affect payments. However, the invoices related to goods and materials have a larger effect on payment time than investment. These results show areas where institutions might be able to improve in order to reduce delays in payments.

Cost estimation

According to our estimates, between 2011 and 2017 the total cost of late payments (after 30 days) was of USD 81.07 million (0.28% of 2017 nominal GDP), meaning that if this new deadline is established and the institutions pay on time, costs could be reduced by this amount. Considering payments are overdue if they are paid in more than 45 days, the cost was of USD 61.12 million

(0.21% of GDP), and in the case of a 60-day deadline the cost was of USD 47.11 million (0.16% of GDP). Moreover, the cost of the total payment duration was of USD 142.29 million (0.48% of 2017 nominal GDP).

Finally, the cost of delaying the invoice stage in more than 15 days is USD 68.63 (0.23% of GDP). As explained in the descriptive results, the invoice stage is the one that accounts for most of the delay, and thus has the biggest cost of the different stages of the process. The complete results are shown in Table 7.

Table 7. Cost estimation of late payments 2011–2017 (full dataset)

Deadline	Cost in millions of dollars	Percentage of 2017 GDP
30 days	81.07	0.28%
45 days	61.12	0.21%
60 days	47.10	0.16%
15 days (invoice stage)	68.63	0.23%
Total cost of payment duration	142.29	0.48%

The results can be interpreted as the costs private providers had to incur in order to face the liquidity problem caused by payment delays. Having longer deadlines, such as 60 days as stated in national regulations, could benefit public institutions since they have to recognise less interests to providers if the payments are delayed and claims are presented. However, this has a negative effect on contractors and at an aggregated level, since they have to find resources to cover their financial needs for the whole payment period, not only after the 60-day deadline. Moreover, since providers know there are delays in payment times, they might internalise the financial cost in the cost structure of the contract, which translates into higher prices for the public administration.

A possible solution to this problem could be to reduce payment deadlines in regulations, to force institutions to accelerate their payment processes and thus reduce the aggregated financial cost. For instance, if the new deadline is set in 30 or 45 days, costs could be reduced by USD 81 million and USD 61 million, respectively. Nevertheless, this must be accompanied by better and more efficient practices inside the procuring entities since, if only the legislation is changed but not the internal processes, this may imply that the institutions might have to recognise more interests for late payments to suppliers. For instance, an evaluation of the European Union Late Payment Directive in 2014, concluded that a new regulation that forced institutions to pay invoices in 30 days or less had not greatly reduced payment times in most of the European countries, but stakeholders argued that legislation had obliged governments to act and improve their payment practices (Valcani Vicari Associati et al. 2015).

When analysed by year, the cost has risen since 2011, but remained relatively constant after 2016. The difference between 2011 and other years is a result of the change in duration in the different stages of the process, as explained in the descriptive results. As a percentage of GDP, the total cost of late payments was of 0.004% of GDP in 2011 and then rose to 0.058% of GDP in 2015 (Figure 14). When considering the total payment duration cost, the number rises to 0.09% of GDP in 2016. In comparison to previous studies about the aggregate cost of late payments in other countries in Europe, Paraguay has a similar cost to countries like Italy or Spain, with a cost close to 1% of annual GDP (Connell 2014). Moreover, the cost is higher in the first quarter of the year and reduces significantly in the last quarter, showing that seasonal trends in payments can increase the costs at specific periods of the year.

Figure 13. Cost of delayed payments by year

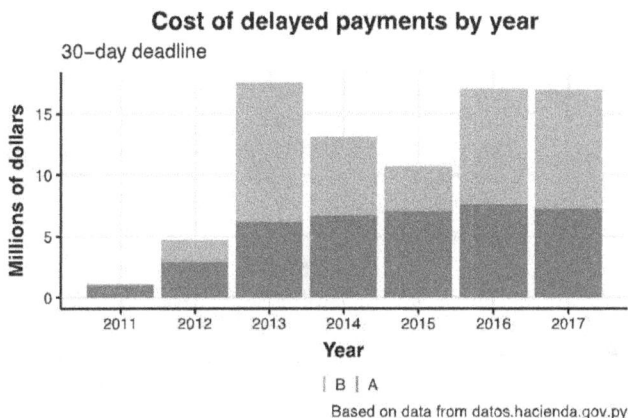

Cost of delayed payments by year
30–day deadline

Millions of dollars / Year

| B | A

Based on data from datos.hacienda.gov.py

Figure 14. Cost as a percentage of GDP

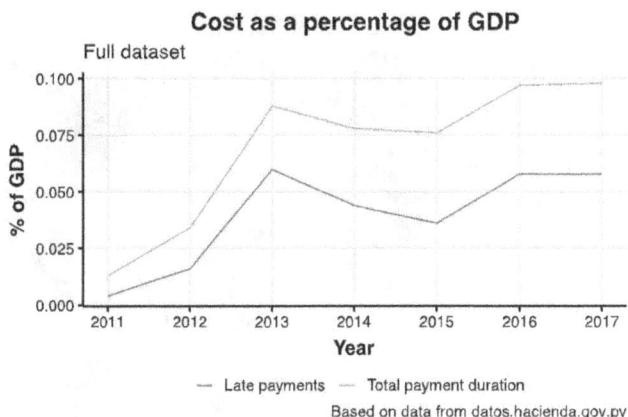

Cost as a percentage of GDP
Full dataset

% of GDP / Year

— Late payments   — Total payment duration

Based on data from datos.hacienda.gov.py

As expected, the larger cost is concentrated in the first stage of the process which accounts for USD 95.8 million or 67% of the total payment duration cost. Changes can be made in this stage in order to improve the process. For instance, if a 15-day deadline is established in this step to encourage institutions to accelerate their payments, the costs could be reduced by 48%.

Figure 15. Total cost by stages

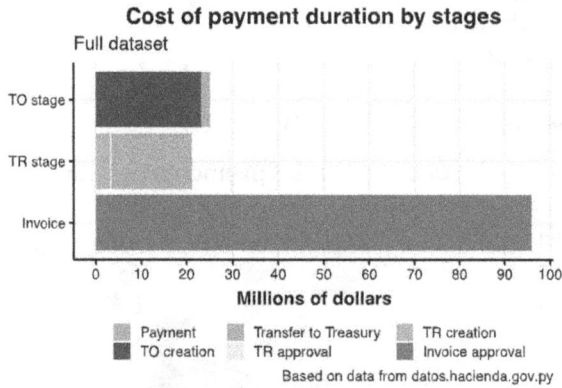

Cost of payment duration by stages

Based on data from datos.hacienda.gov.py

Finally, three entities concentrate 73.9% of the total payment duration cost. The Ministry of Public Works accumulates 37% of the total cost, the Ministry of Health 30% and the Ministry of Education 5.9%. This means that procuring payment practices must be improved in these institutions in order to obtain the biggest savings.

Figure 16. Cost of delayed payments by institution

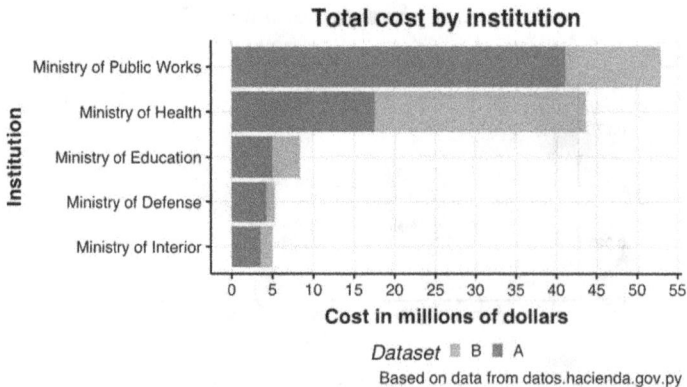

Total cost by institution

Based on data from datos.hacienda.gov.py

130

Conclusion

This work shows how to calculate the cost of the duration and delay of payments and the variables that affect payment time, using detailed open contracting data. We found that the total cost of the payment duration in Paraguay, between 2011 and 2017, was of USD 142.29 million, equivalent to 0.48% of Paraguay's 2017 Gross Domestic Product. Considering a 30-day deadline as an acceptable payment time, the cost of late payments was of USD 81.07 million. In general, procuring entities take 55 days on average to pay invoices, while the international optimal deadlines are established in 30 days. For some institutions, the duration can extend for more than a year, which shows there are inefficiencies in the procurement payment process.

Moreover, two ministries, Health and Public Works, concentrate 67% of the total cost, so the corrective efforts should be concentrated in these entities. There is also evidence that the funding source, the type of purchase, the size of the institution and its budget execution affect payment duration times. Besides, the first step of the payment procedure (invoice stage), which is not regulated, is the one that takes longer. Our recommendation would be to revise the procurement practices at this stage in order to shorten the delays. If a deadline of 15 days is met, costs could be reduced by 48%. Moreover, if the total payment deadline is established in 45 or 30 days since the invoice issuance, the total payment duration cost could be reduced by 42% or 52%, respectively.

We conclude that there are two main areas of improvement in the payment process. First, regulate the first stage and establish a deadline to force institutions to modify their practices to load invoices into the system more quickly. This is particularly important for invoices paid with institutional and public credit funds, since there is no justification to delay the process if there are resources available to pay. In the case of bills paid with Treasury funds, our recommendation would be to modify the steps in the process, so that invoices can be loaded into the system without a cash plan. This can help the Treasury identify which payments are coming due and distribute the funds accordingly.

Moreover, this work demonstrates how public procurement open data can be analysed to generate high value insights for the public administration, in order to improve contracting practices and save public funds. Our methodology can be implemented in any country that publishes payment and contracting data in open formats, and aims to serve as an example of how institutional efforts to publish detailed open data about all the steps of the procurement process, can pay off when valuable insights are derived from its analysis.

Our study faced some limitations regarding mainly the availability of data. It would be useful to have more disaggregated data about the providers in order to determine whether the delays are affecting more small, medium enterprises or big companies. Even though the cost can be higher to bigger providers that obtain larger contracts, for smaller enterprises it could be harder to access credit

131

or to cover the lack of liquidity caused by the delay. In addition, not having information about the characteristics of the firm (size, sector) can result in an underestimation of the cost, since the interest rates can be different. Having open data about beneficial ownership could help expand the analysis to determine if there are clusters of providers that are more affected by the delays, and could also be useful to estimate the effect of late payments on the exit rate of firms. In addition, the analysis could be extended if payment information of other institutions in the public sector published payment data in open formats.

Finally, this investigation could be extended with a dataset of the cash plan, to compare the availability of resources with the payment information, in order to determine more efficient ways to redistribute the resources and thus reduce payment duration.

REFERENCES

Balaeva O & Yakovlev A (2017) Estimation of costs in the Russian public procurement system. *International Journal of Procurement Management* 10(1). DOI 10.1504/IJPM.2017.10000827

Cameron C & Trivedi P (2005) *Microeconometrics: Methods and Applications*. Cambridge: Cambridge University Press

Checherita-Westphal C, Klemm A & Viefers P (2016) Governments' payment discipline: The macroeconomic impact of public payment delays and arrears. *Journal of Macroeconomics* 47: 147–165. DOI 10.1016/j.jmacro.2015.12.003

Connell W (2014) The economic impact of late payments. Directorate General Economic and Financial Affairs (DG ECFIN) Economic Paper No. 531. Brussels: European Commission

Diamond J & Schiller C (1987) Government arrears in fiscal adjustment programs. *FinanzArchiv / Public Finance Analysis* 45(2): 229–259

Fiordelisi F, Mare D, Radic N & Ricci O (2012) *Government late payments: The effect on the Italian economy – I ritardi di pagamento della Pubblica Amministrazione*. Financial Intermediation Network of European Studies (FINEST) Report No. 1. https://assifact.it/wp-content/uploads/2016/12/FINEST-Report_24042012.pdf

Flynn S & Pessoa M (2014) Prevention and management of government expenditure arrears. IMF Technical Notes and Manuals 14/03. Washington, DC: IMF

Giussani B, Guardiola U & Ospina J (2016) Evaluación PEFA de la Gestión de Finanzas Públicas en Paraguay. PEFA. https://pefa.org/assessments/paraguay-2016

Gori G, Lattarulo P & Mariani M (2017) Understanding the procurement performance of local governments: A duration analysis of public works. *Environment and Planning C: Politics and Space* 35(5): 809–827. DOI https://doi.org/10.1177/0263774X16680109

Guccio C, Pignataro G & Rizzo I (2012) Measuring the efficient management of public works contracts: A non-parametric approach. *Journal of Public Procurement* 12(4): 528–546

Guccio C, Pignataro G & Rizzo I (2014) Evaluating the performance of public procurement contracts for cultural heritage conservation works in Italy. *Journal of Cultural Economics* 38(1): 43–70. DOI https://doi.org/10.1007/s10824-012-9194-2

Marchessault L (2013) Open contracting: New frontier for transparency and accountability. World Bank Institute. https://www.open-contracting.org/resources/open-contracting-a-new-frontier-for-transparency-and-accountability/

Open Contracting Partnership (2017) *Annual Report 2017.* https://www.open-contracting. org/resources/annual-report-2017-serving-transparency-change-public-contracting/

Ramos A (1998) Government Expenditure Arrears: Securitization and Other Solutions. *Working paper of the International Monetary Fund* No. 98/70. https://www.imf. org/en/Publications/WP/Issues/2016/12/30/Government-Expenditure-Arrears-Securitization-and-Other-Solutions-2596

Smirnov J (2016) Modelling late invoice payment times using survival analysis and random forests techniques. MSc thesis, Faculty of Science & Technology, University of Tartu

Strand I, Ramada P, Canton E, Muller P, Devnani S, Bas PD & Dvergsdal K (2011) *Public procurement in Europe: Cost and effectiveness.* A Study on Procurement Regulation Prepared for the European Commission. Brussels: London Economics, PricewaterhouseCoopers and Ecorys

Valcani Vicari Associati, Technopolis Group & Ernst & Young (2015) *Ex-post evaluation of late payment directive.* Brussels: European Commission. https://publications.europa. eu/en/publication-detail/-/publication/400ecc74-9a54-11e5-b3b7-01aa75ed71a1

World Bank (2008) *Paraguay – Integrated fiduciary assessment (English).* Washington DC: World Bank Group http://documents.worldbank.org/curated/ en/620621468070136312/Paraguay-Integrated-fiduciary-assessment

World Bank (2017) *Benchmarking public procurement: Assessing Public Procurement regulatory systems in 180 economies.* Washington DC: World Bank Group

7.

Connecting flows and places: Flows of (open) data to, from and within hyperlocal communities in Tanzania

François van Schalkwyk

This chapter presents findings from research on the flow of data in the United Republic of Tanzania, and specifically on the flow of data to and from the hyperlocal level in the health system of that country. The research was undertaken to provide open data initiatives wishing to stimulate the evolution of a data system with empirically-based evidence on the blockages, breaks and connectors in the flow of data in such systems, as well as the possible role of open data in creating new flows to communities at the hyperlocal level.

Understanding data flows in Africa

African countries have pledged to achieve sustainable development and inclusive growth by adopting the 2030 Agenda for Sustainable Development and the Agenda 2063. At the national level, this can be seen in long-term national development plans and numerous legal, legislative and policy reforms aimed at improving the quality, timeliness, relevance, availability and accessibility of data. At the time of writing, Ethiopia, Ghana, Mauritius, Rwanda, Sierra Leone, Tanzania and Tunisia had published data policies, or were in the process of doing so, and there were at least 12 separate national government-led open data initiatives on the continent. But data alone is not enough to drive development – equally important are the actors, infrastructures, technologies and emergent social dynamics, including the distribution of power, that shape the flow of data, in so doing connecting data to the policy- and decision-makers instrumental in the allocation and distribution of resources in pursuit of equitable human development.

To illustrate, data are routinely collected at the local level by local governments across the globe. These data are fed into state, national or federal management

135

information systems for planning purposes which ultimately provide the basis for determining the distribution of resources (political agendas and corrupt practices notwithstanding). At the same time, data feed indicators of a nation's financial health, including, for example, its economic growth and the investment climate which, in turn, rely on data describing the social conditions and well-being of defined populations. This is vital data that, when circulating in global financial networks, determines the ability of the state to access (and provide) financial instruments in the globalised financial system. This is not only about access to global financial networks, but about nation states becoming inexorably integrated into the global financial system. According to Castells (2017), three major processes are changing the coordinates of the global political economy, one of which is that global financial markets are increasingly the core of national and international economies.

However, the primary data and those who inhabit the local spaces from which primary data originate, typically remain disconnected (Mori et al. 2014; Wickremasinghe et al. 2016) from decision-making. Custer et al. (2018) argue that improvements in sub-national data collection will bring data closer to place-based decision-making. According to Castells (2009, 2010) the internet and communication technologies that have been developed to exploit real-time connectivity on a global scale have had profound effects on society. This is most evident in how society is being reorganised according to the programmes of global networks and the effects of this restructuring on the development of society (Castells & Himanen 2014). There are legitimate concerns that, rather than making possible a more equitable distribution of resources, the network society is one in which exclusion is structurally manifest and the gap between the rich and the poor, the powerful and the marginalised, the metropolitan and rural, is certain to widen (Castells & Himanen 2014; Ravallion 2016).

At the risk of oversimplification, Castells (2010) explains this structural binary divide as being attributable to a particular condition of the network society: the cleavage between two spaces – the space of flows and the space of places. The space of flows describes that placeless space where information is exchanged in real-time across global networks. The space of places describes the local, physical spaces in which social actors live and breathe; where they seek meaning and define who they are. The disconnect between these two spaces in the network society provides a useful framework for understanding the breakages and asymmetries in the flow of data between local communities, their governments and global networks. The approach also offers a more nuanced account of the contingent relationship between data and the spaces in which data are situated at any given time.

Open data could play a role in ameliorating the exclusionary effects of networks by democratising access and the use of data as a highly prized commodity in global networks (Van Schalkwyk & Cañares 2020). More equitable access, in turn, makes possible, in theory at least, the reprogramming of networks by certain strategically located actors in networks, that is, switchers and programmers (Castells 2009, 2010). Switchers connect networks, programmers set the overarching logic of the network. In plain terms, open data can mobilise social movements to effect change in networks programmed according to logics that will not deliver equal and sustainable societal benefits. Governments themselves, as programmers, may take the lead to leverage open data for more equitable outcomes. In the health sector, for example, governments in various parts of the world are seeking to unlock the potential of open data to improve the quality of care, lower healthcare costs and facilitate patient choice (Scrollini 2017; Van Schalkwyk et al. 2017; Verhulst et al. 2014).

To conclude, there remains a lacuna when it comes to an understanding of the flows of data, particularly as it relates to the local level. This research therefore poses two questions: 'How does data flow in a specific data system?' and 'What is the contribution of open data in stimulating data flows in the data system?'

Analysing data flows

Data flows are best understood with reference to the movement and exchange of raw and processed data between humans and machines in complex socio-technical systems. This approach is consistent with those adopted by others (see, for example, Bates et al. 2016; Kitchin & Lauriault 2014) in its attempt to illuminate 'the concrete ways in which evolving socio-cultural values and material factors cohere over time to create the socio-material conditions that frame activities of data production, processing and distribution and resultantly influence the form and use of data' (Bates et al. 2016).

In this chapter, the use of data flows is preferred to 'data journeys' as proposed by Bates et al. (2016). This is because the notion of a journey implies a degree of agency to data, a purpose. While it is true that data is not necessarily neutral and is subject to manipulation which, in turn, can have profound effects on decision-making, both its neutrality and its purpose are compromised and determined by the actions of humans (or possibly machines). Data has no inherent volition or velocity; data are socially constituted and its emergent materiality is socially dependent and technologically enabled. Furthermore, the concept of flows, like journeys, does convey important properties; properties that indicate how data moves through complex socio-technical systems. Flows may be rapid, slow or non-existent (Van Schalkwyk et al. 2016); flows may be channelled, redirected or divided; and flows may be fragmented, partial or disconnected within and between systems.

Data flows imply a flow between, from one point in time and space to another point. The situatedness and movement of data has led scholars to draw attention to the importance of data infrastructures (Dodds & Wells 2019). Rather than existing in isolation, data infrastructures are connected and arranged into data assemblages (Kitchin & Lauriault 2014). Assemblages shape data and their flows are, in turn, shaped by both the data and their socio-materiality (Bates et al. 2016; Kitchin & Lauriault 2014). The limitation of data assemblages is that while the approach acknowledges the social contingency of any assemblage, it tends towards a focus on connections within assemblages at the expense of connections between assemblages in broader networks as well as the connections between those networks. As a consequence, the effects of 'meta-assemblages' and, more importantly, of global networks acting on one another is lost.

This limitation is illustrated by the often-used analogy of data infrastructures resembling road networks (see, for example, Dodds & Wells 2019). The rules of the road are typically defined at the national rather than the global level; while roads connect physically across national borders, it is not possible to connect across large bodies of water and there is competition for road space resulting in traffic congestion. Data does not have to contend with these impediments by virtue of its non-rivalrous, digital nature and a global, fully interconnected network (the internet) for its movement.

The arrangement of actors and the dynamic conditions in which data production, processing, distribution and use occur, have led other scholars to refer to assemblages as 'ecosystems' (Harrison et al. 2012; Heimstädt et al. 2014; Van Schalkwyk et al. 2016). Ecosystems consist of mutually interacting organisms; are complex in their arrangement; characterised by the interdependency of and between organisms and resources; are dynamic rather than static (seeking equilibrium through motion rather than stasis); populated by keystone species that play a critical role in facilitating exchange in the ecosystem thereby ensuring dynamism and constant movement; movement tends to be cyclical and reinforcing, making the system resilient (adaptable and restorative) but ecosystems are also vulnerable to exogenous forces which may disrupt or destroy the ecosystem (Van Schalkwyk et al. 2016).

While both data assemblages and ecosystems are clearly complex, they can be mapped by identifying the components of the system and the relationships between those components. Components include human and non-human agents, as well as the routines, rules and norms that emerge from their interaction.

The benefit of the ecosystem approach is that it fits more comfortably with a network understanding of society as described in the introductory section to this chapter. A network approach situates data, its assemblages and flows within a broader socio-political framework that accounts for the social forces that act upon data. The approach brings to the surface not only an understanding of the component parts, movement and endogenous forces acting on data within assemblages but also of the exogenous and divergent social forces that shape those assemblages. Such an understanding can account for friction in the flows of data (Bates et al. 2016) between those operating in global networks and those local communities external to those networks as distinctive sites of practice subject to different logics or programmes. Kitchin and Lauriault (2014: n.p.) write that 'databases and repositories are expressions of knowledge/power, shaping what

138

questions can be asked, how they are asked, how they are answered, how the answers are deployed, and who can ask them'. An acknowledgement of the inescapable effects of how power is distributed and exercised in socio-technical systems (or networks) is critical; the data flows approach adopted in this chapter accounts for such power dynamics but takes the analysis beyond databases and repositories into the complexity of a networked social world defined, in part, by the cleavage between non-material spaces of flows and the local place-based spaces inhabited by local communities.

Methods

The approach of this study is informed by and consistent with critical data studies, that is, studies that seek to develop critical, qualitative methodologies to enhance our understanding of the place of data in society (Kitchin & Lauriault 2014). It also seeks to contribute to previous empirical studies on the subject, for which there remains a need (Bates et al. 2016).

This research formed part of the Data Zetu[1] project which sought to amplify citizens' voices through data. The Data Zetu project was located in Tanzania and was funded and implemented in partnership by the United States President's Emergency Plan for AIDS Relief (PEPFAR) and the Millennium Challenge Corporation (MCC) as part of the Data Collaboratives for Local Impact programme (DCLI).[2] The scope of the research was therefore defined by the activities of the project, both in terms of sectors and geography.

From the regions in which the project was active, four wards in the Kyela district in the Mbeya region located in south-western Tanzania were selected as the study site. Selection was based on the remoteness of the district from any urban centres more likely to be connected to global networks and by the degree of access to key informants in the districts. The wards in Kyela district constitute the hyperlocal level in this context; adding a more fine-grained level of analysis to those typically used (i.e. supranational, national and sub-national).

Information on the available datasets in the ecosystem was collected in the first instance by means of desk research and in consultation with the Tanzania Data Lab (dLab) as it works closely with both the national bureau of statistics (NBS) and civil society organisations (CSOs) in the health sector in Tanzania. By conducting interviews, searching the media and consulting other data mapping efforts, 37 health-related data sources were identified and coded for: name (of the dataset); collection type (e.g. database; dashboard; document repository); types of data (e.g. health; demographics; education); keywords (describing the data, e.g. mortality, malaria, TB); collection (method of collecting the data, e.g. routine, survey, third-party); data owners (primary = owner of the source

---

1   https://datazetu.or.tz/
2   https://www.mcc.gov/initiatives/initiative/mcc-pepfar-partnership

data; secondary = owner or host of the platform or medium for accessing the data); data source (from whom is the data collected, e.g. district office, general population, health facility); level (e.g. supranational, national, regional); formats (e.g. excel, PDF, csv); most recent data available (year); available online (yes/no); accessible (yes/no); open data (yes/no) and URL (to dataset or to page where data can be found). This list was not meant to be exhaustive but rather to represent a sample of the types of Tanzanian health data available and, important in the context of this study, the levels at which the data are made available.

Once the identification of health datasets was complete, the flow of data was traced from the local to the national and/or supranational levels and vice versa, using the list of data sources. Further unstructured, in-person interviews conducted in December 2017 and March 2018 verified mapped flows, identified data and data sources previously not identified and provided insights into context-specific data practices.

To map more precisely data flows and practice at the hyperlocal level, fieldwork was conducted in Kyela district in August 2018. A structured questionnaire was administered to government health facility supervisors at 31 health facilities in the Kyela district, to the district executive director and to the district medical officer (DMO). The data were captured in a quick-tap survey application using hand-held devices and exported to Microsoft Excel for cleaning and analysis.

Findings

The findings of the data flow mapping exercise for the health data in Tanzania is presented graphically in Figure 1. The ministry of health (MH) is located at the top of the diagram and represents the national level at which health data is integrated, controlled and analysed. At the bottom of the diagram is the hyperlocal level, represented in this case by the communities and health facilities of Kyela district. On each side of the diagram are the other actors identified in the data system, loosely grouped into CSOs to the bottom left and research organisations to the bottom right of the diagram.

In the sections that follow, the flow of data in the system is discussed. It is acknowledged that mapping a complex system graphically will always fall short of the reality which it attempts to represent. Nevertheless, it is hoped that even an oversimplified representation of reality can provide some useful insights into how data flows in the health data system in Tanzania.

Vertical data flows from the hyperlocal level to the national level

Vertical data flows describe top-down and bottom-up flows. In this section, the focus is on bottom-up flows, that is, from the hyperlocal level to the national level. A later section deals with return or top-down flows.

Figure 1 shows that at the most granular level, health data is collected at health facilities located in districts. All public and private health facilities (hospitals, dispensaries, clinics and health centres) capture patient and operational data and submit these to the district medical officer's (DMO) office. Data on deaths, births, outpatients, inpatients, ante- and postnatal care, vaccinations and HIV/Aids treatment are just some of the data types that must be collected at the facility level.

Much of the data capturing is still done in paper format using prescribed printed forms provided by the MH. Data are captured in counter books before being transcribed to the MH's standardised forms. It was found that 19% of health facilities capture data only in paper format with the remaining 81% using both paper and electronic data capturing methods. Data are typically converted into electronic format for submission to the DMO, although some still submit data to the DMO only in paper format. This finding confirms previous research that reported that as much as 30% of facility-level data is submitted to the district medical office in paper format (Bhatia et al. 2016).

At the DMO, health facilities data are captured in the central health management information system software, the District Health Information System (dhis2),[3] by DMO office staff. This constitutes a second instance of data capturing.

In addition to health data received from health facilities, DMOs capture in dhis2 other data specific to their district (e.g. demographic data). These data are required for the production of health indicators for monitoring and evaluation purposes at the district level and are sourced from other government departments and agencies such as the national bureau of statistics. It was not established how officials in the DMO's office access and capture non-health data obtained from NBS.

Data on health commodities (i.e. medicines) are captured in the national electronic logistics management information system (eLMIS) and not in dhis2. In Tanzania, the eLMIS collects data from more than 6 000 service delivery points (USAID 2015). Dhis2 is linked to eLMIS and extracts selected data from that database (e.g. on tracer medicines) to produce certain health indicators. More than one interviewee commented that while the design of the eLMIS system looks good on paper, there are concerns about the comprehensiveness and quality of the data captured in eLMIS.

CSOs and other project-based initiates also participate in the collection of routine data at health facilities. These data are also sent to the DMO for capturing into dhis2. CSOs are therefore active participants in the health data system but do not capture health data directly into any of the central health data systems administered by the government.

---

3   DHIS2 (District Health Information System) is used in more than 60 countries and is an open source software platform for reporting, analysis and dissemination of data for all health programmes, developed by the Health Information Systems Programme.

Findings show that data do not flow from non-facility sources such as health surveillance sites or outposts to dhis2. Nor are survey or surveillance sentinel data captured in the dhis2 system, either via the DMO or from the National Health Institute for Medical Research or Ifakara Health Institute, the latter being the corner-stone institution for government health research.

Survey data are also collected at the population level in the form of population-based surveys such as the Demographic and Health Survey and the HIV/Aids and Malaria Indicator Survey conducted by the national bureau of statistics; the Population and Housing Census also conducted by NBS; the Demographic Surveillance System, which monitors vital statistics at sentinel sites located at various regions in the country; sample vital registration with verbal autopsy, which operates under the sentinel panel of districts and national vital registration systems and other specific health and health-related research works. Results from surveys are typically presented in report-format to MH and other interested government officials.

Therefore, any medical or health-related event that occurs within communities, and does not involve a health facility, will not be captured in dhis2. Such data are not included in any analysis done using dhis2 data, be it by the MH, the President's Office Regional Administration and Local Government offices (PO-RALG) or the Council Health Management Teams (see below for more detail).

According to a DMO, the '[District Executive] accepted that village leaders can provide accurate data [when] we were distributing nets. They have been insufficient all the time due to underestimating population. After realizing that, we have been working with village leaders in identifying children under five and mothers. This information is always exact' (Bhatia et al. 2016: 12). In this case the medical officer is pointing out the problems of relying on inaccurate population data from the national level when distributing mosquito nets and that the community is able to provide more reliable data to ensure more effective intervention. However, mechanisms for connecting village leaders to health workers are non-functional and therefore do not facilitate the flow of data into the national health system via local health facilities (see also Silaa & Van Schalkwyk 2018).

The mHealth Tanzania Public-Private Partnership is an initiative that makes possible upward flows from isolated or disconnected health workers. The Partnership focuses on addressing ministry-defined public health priorities by supporting solutions that work in concert with initiatives underway at the MH. mHealth initiatives include those that provide direct health communications to citizens via SMS and clinical decision information and reminder services for health workers. These services rely on existing data held by the MH. But health data is also captured by mHealth initiatives by, for example, the Infectious Disease Reporting System (IDRS) which allows health-facility workers to report disease surveillance data by making a free call from the field using any mobile phone. Real-time SMS and email alerts are then generated by the system for follow-up and action.

Horizontal data flows: National level

Horizontal data flows describe those data flows between actors operating at the same level in the data ecosystem. Figure 1 shows intra-governmental data flows between MH and the NBS as well as the flows within divisions of MH.

Data flows from NBS to MH are either by request (usually for microdata) or by presentation (aggregated summary data), typically following the completion of one of the national surveys. Data are presented by NBS as aggregated data and indicators in reports made available in print and PDF formats. Data requests are usually activated and responded to at a high level across government agencies. For example, data requests to NBS often originate in parliament. In such cases, data requests are made by the permanent secretary in the President's Office to the Director of NBS.

Data flows within MH show evidence of increased levels of coordination and integration. For example, collection of HIV/Aids data has been integrated into dhis2. According to the National Aids Control Programme's 'National Guidelines on HIV and Aids Data Management' report, data is collected at facility-level using the CTC2 database and dhis2 is the main repository for HIV and Aids data (MOHCDGEC 2017). Aggregated quarterly reports from CTC2 are captured in dhis2 while granular patient data remains in the CTC2 system. The HIV/Aids client records database has scaled to more than 900 facilities, improving record keeping for tracking of HIV/Aids clients (PATH 2017). The TB/Polio database remains separate but shares architecture with dhis2. The eLMIS database also remains separate from dhis2, but dhis2 draws on data from that database in order to produce key health indicators.

Return, vertical data flows: National level to hyperlocal level

According to the World Health Organization (WHO):

> Access to health information [in Tanzania] for all levels from the general community up to decision-makers, and the utilisation of the generated health information, have been inadequate and this is a major challenge to the health system. There are no standardised methodologies in place to ensure appropriate information is channeled to the right person at the right time and for the right purpose. Thus, there has been poor utilisation of the available information for knowledge strengthening and supportive evidence for decision-making. (n.d.)

District-level health management teams have access to dhis2 data and these data are accessed via reports generated by the dhis2 system. Bhatia et al. (2016) report, however, that analysis and planning at the district level cannot rely on dhis2 data alone and requires deeper analysis. The fact that community and

143

other data sources are not available to district officials may hamper such analysis. For example, data from sentinel surveillance (including local data on births, deaths and burden of disease) are not used to calibrate and validate findings from facility-based data. And the fact that dhis2 does not include district-specific indicators (i.e. indicators relevant to specific district health needs and not included in the dhis2 indicators that are determined by national priorities), places further constraints on the availability and use of fine-grained, context-specific health data. A case in point in Kyela district are the seasonal cholera outbreaks that are not prioritised to the same extent as HIV/Aids and TB at the national and supranational levels. Data on cholera outbreaks are collected by several health facilities in Kyela district but there are no indicators for cholera on the national portals (Silaa & Van Schalkwyk 2018).

Communities do not have any direct access to health data held by facilities nor do they have indirect access via intermediaries such as faith-based organisations or CSOs.

Horizontal data flows: hyperlocal level

All dhis2 district-level health data is centralised by the District Medical Office and health commodities data is centralised by the national Medicine Stores Department. While they are the primary source of the national government's health data, health facilities can only access their own data on the national health management information system (dhis2) and on other relevant systems (such as eLMIS). They are therefore unable to access electronically the health data collected by neighbouring health facilities in the same district.

Limited horizontal access to health data at the facility level should not be interpreted as a lack of information sharing: 29 (90%) of facilities indicated that they are able to access health data from other facilities. Of those 29 health facilities, 94% indicated that they request data by telephone. In most cases (65%) requests are made only by telephone while in other instances, telephone calls are combined with a visit to the facility (10%) or with sending a WhatsApp message (19%). A small number of respondents (6%) indicated they make exclusive use of WhatsApp to request data from other health facilities. Limited, informal data sharing takes place in the absence of formal mechanisms to facilitate data sharing.

Research has also shown that CSOs are not publishing or sharing their health-related data (Tunga & Mushi 2016) with other actors in the health system and this was confirmed in the interviews conducted. The sharing of health data among CSOs and between CSOs and the general public is poor. Mostly, data are used internally to produce reports, briefs and other outputs in support of advocacy work. Consequently, data is shared by disseminating printed materials or by making presentations to relevant stakeholders.

Figure 1. Data flows in the Tanzania health system

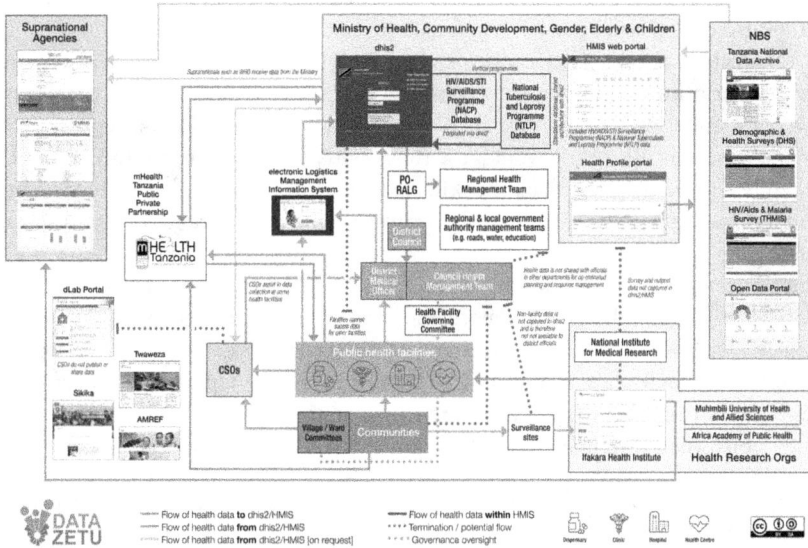

Feedback loops, data quality and terminations

In its 2018 policy, the MH states that the objectives for establishing a monitoring and evaluation system for the health sector in Tanzania include:

- Providing a mechanism for feedback to update the information system;
- Providing reports to all stakeholders with necessary and sufficient data (MOHCDGEC 2018).

Figure 1 shows that only one feedback loop is in operation – the supply of facility data by the district medical office to dhis2 and the use of the same data by district health planning officials, including the Council Health Management Team, for planning, response and resource allocation to facilities that provided the data.

Feedback loops are organic, social arrangements that depend on sanctions or rewards to operate. Sanctions assume the form of either explicit rules or institutionalised norms. Rewards exist in the form of incentives for the initiator of the feedback loop and for all subsequent data contributors and users. Incentives may be either material or non-material. Findings suggest that health officials at the facility level are motivated to submit health data on time, most likely because timely submission is a key indicator in their performance evaluations. Their motivation therefore appears to be driven by the threat of sanctions rather than on incentives that could, for example, be linked to the quality of the data submitted (see also Bhatia et al. 2016; Sato et al. 2017). On this basis, upward flows appear to be efficiency- rather than quality-driven and this has serious implications for the usefulness of the data for planning and resource allocation in return downward flows of health data in Tanzania.

Interviewees expressed concerns about the quality of the data transcribed as transcription is often done by nurses or administrators with poor data skills who are under pressure to submit the data on time. Kimunai (n.d.) reports that some health facility staff would ask those with data skills to capture data on their behalf because they lack the skills to do so. 'Data cooking' was cited by one interviewee as being common, as data capturers make up data to compensate for data gaps in the health facilities' records. Sato et al. (2017) measured aspects of motivation among health workers in rural Tanzania and found that having a clear job description is the greatest motivation for health workers in executing their functions. It follows that if nurses and other health workers whose description does not include data transcription and capturing for reporting purposes (as opposed for patient management) are expected to undertake these data-related functions, then their motivation for completing these functions accurately, will be low.

Dhis2 officials indicated that data quality tools have been made available to health officials to check the quality of the data submitted to and by the DMO and that the MH has been rolling out data training at the facility level to improve skills and capacity in data capturing. At the DMO, a Health Management Information System (HMIS) coordinator validates the data in the system based on observed data anomalies in which case the health facility is contacted for verification. However, respondents indicated that corrected data may or may not be updated in dhis2 because after data submission deadline, making changes to the data in the system raises 'a lot of issues'. The implication of this reluctance to update incorrect data is that the correct data will be held at the DMO but will not be in dhis2. It was also indicated that some data go through more thorough checking processes because donor funding is available. This means that some types of health data (e.g. HIV-related data) receive more attention based on the funding priorities of supranational donors.

While training may help to prevent errors and tools may flag obvious data errors, it is not clear how manufactured data can be prevented and detected when the incentives for accurate data collection and for capturing corrections are absent at the hyperlocal level. This raises serious concerns about the quality of the data in dhis2 used to conduct modelling, analysis and, ultimately, to make decisions about the allocation of health resources.

In many other instances, Figure 1 shows terminations in the flow of data, thus precluding the creation of feedback loops. Boerma (2013) argues that districts require health information systems that draw from multiple sources. To some extent, integration of data sources has taken place at the national level, but non-facility health data from surveillance sites, from the community and from CSOs do not flow into the health information system. Thus, there is no formalised system for DMOs and health teams to access non-facility health data. And, as a consequence, there are no systematic local feedback loops between health officials and these sources of health data. As decentralisation in Tanzania continues, local decision-makers will have more opportunities to use results data for local decisions.

At the same time, CSOs are either loath or ill-equipped to share health data. Findings show that research institutes appear more inclined to share data with international donors than with either the national health ministry or local health officials. Health facilities in the districts are not able to access each other's administrative and performance data via dhis2. If they were able to do so, this could improve district-level coordination in cooperation with the DMO and other stakeholders active at the hyperlocal level.

Open data

Several sources of open health data are available in the Tanzania health data system, although they vary in the extent to which they meet some of characteristics often prescribed for open data such as being published under an open license, being machine readable and being available in bulk. Nevertheless, these datasets are all online and accessible without restriction.

Open data made available by the government comes from three main sources. The first is data published by the MH predominantly from its health management information system, dhis2, on two publicly accessible data portals: (1) the HMIS web portal[4] and (2) the Health Profile portal.[5] The HMIS portal makes available indicator and raw data primarily on diseases but also includes demographic data (e.g. population by age), data on health personnel (e.g. health workers by cadre) and the completeness of health data collection across the system. The HMIS portal includes data from two non-facility sources: the HIV/Aids/STI Surveillance Program (NACP) and the National Tuberculosis and Leprosy Program (NTLP). The NACP's data has been integrated into the dhis2 while the NTLP remains a separate database from which dhis2 extracts relevant data for producing indicators. The Health Profile portal's focus is on data and indicators related to health service delivery, including resourcing, medicines and infrastructure. Both publish data disaggregated to the district level.

The second source of open data is from the NBS. Data is published on the Tanzania National Data Archive and includes data from the health-related surveys conducted by NBS. The Tanzania Open Data Portal also falls under NBS and contains 12 health-related datasets mostly uploaded by the MH.

A third source of open health data is the data published by the Ifakara Health Institute. At the time of writing only 8 of 109 datasets listed in its central data

---

4   https://hmisportal.moh.go.tz/hmisportal/#/
5   https://hmisportal.moh.go.tz/dhpportal/#/home

repository were either available as 'Direct data access' or as 'Public use data files'.

Many of the open data originating from these government sources are republished in supranational open data sources such as those of the World Bank, WHO, UNICEF and many of the donor organisations such as USAID and the Gates Foundation that support the collection of survey data or provide funding for health research in the country.

Both communities and health workers at the facility level can, provided they have the means and the skills to do so, access the HMIS public web portals that provide periodic data at the district level, but not at the level of facility. Using the HMIS web portals requires more than literacy; the portal describes data on the portal using sector-specific jargon and acronyms making it difficult for non-specialists to find and understand the data presented. The portals were also found to be slow and unreliable and contained several incomplete data fields.

Findings from the interviews conducted at the health facilities show that 23 (74%) of the health facility supervisors interviewed were not aware of the government's open health data portals and only 4 (13%) had made use of the portal in one way or another. Reasons given for using the health portals included accessing health guideline updates, familiarisation with 'hot cases' (disease outbreaks), for reporting and medicine requests and for personal reasons. When asked whether they had ever used other, non-MH data sources (e.g. World Health Organisation, UNICEF), 3 (10%) respondents indicated that they had used non-government sources, mainly to find new information pertaining to vaccination and disease management.

CSOs indicate that they struggle to access government health data, both at the national level and at the district level. One interviewee indicated that while government welcomes the health system monitoring function performed by her organisation, government also responds to requests for data by saying that it does not want 'our own data to be used to kill us'. There are, however, examples of district health management teams and facilities responding positively to requests for data and express willingness to engage with social accountability monitoring initiatives.

The dLab has been working with health CSOs on the publication of open health data and has made available a data portal to facilitate data publication. To date, hardly any CSOs have made use of the portal. CSOs that receive funding from donors who mandate data sharing as a condition of grants awarded, are publishing anonymised datasets either on the data repository of the funder (e.g. Wellcome Trust and The Bill & Melinda Gates Foundation) or on an approved data repository (e.g. Harvard Dataverse, Figshare).

The MH appears to be aware of the disconnect between research and decision-makers. It has proposed in its most recent policy document (MOHCDGEC 2018) the activation of a National Health Research Forum as a platform for knowledge exchange that will surface regional health priorities. The MH also commits to an open access policy for research to make research more widely accessible. These are both positive steps in integrating research data into the health information system. However, other than to reference a 'Data Dissemination and Use Strategy', the policy remains silent on exactly how research data will be connected to and/or integrated into the health management information system.

Research data and other non-facility health data are not incorporated into the national health management information system. There is also no mechanism for citizen-generated data, such as community-mapped health information, to be captured in any of the central health management information systems. Community reporting could play an important role in upward information flows via the IDSR or other mHealth initiatives so that communities join local and community health workers to create a wider network of disease reporters.

Discussion

This study has shown that data flows are complex. While empirically valuable, this finding should come as no surprise given the contingency of data flows on social dynamics, that is, on what Bates et al. (2016) refer to as the 'socio-material' constitution of data flows. Integrating and connecting flows are not purely technical undertakings but require overcoming socially constructed forces shaped by the distribution of power (the political) and by taken for granted and relatively immutable practices (the institutional).

A clear instance of a socially constructed force and its resultant arrangement of data infrastructures is the number of vertical health programmes in Tanzania defined by international donor agencies. Health programmes for diseases such as TB, malaria and HIV/Aids were siloed within MH with the support of particular donors and each programme developed its own data infrastructures to monitor and evaluate progress being made in combating its targeted disease. The findings show the remnants of these external interventions but also the progress being made in breaking down divisions between infrastructures (e.g. linking the TB reporting system to dhis2) in order to improve horizontal data flows.

The findings also show that data collection is highly formalised and centralised via the Health Management Information System's dhis2 software around two points in the data system: at the national level (by the MH) and at the district level (via the DMOs). The practice of capturing health data on paper combined with multiple instances of transcription slow the upward flow of data and indicate problems related to data quality which, in turn, could constrain data use. In general terms, data use is still relatively weak by virtue of the fact that data practice at all levels is driven by reporting requirements set at various levels of government rather than by the value of data for informed planning and decision-making.

Findings show that data flows are more definitively outwards and upwards from local communities and their health facilities than they are inwards or downwards. There are also several outward and inward data flows that terminate. These terminations stagnate the flow of selected health data in the system. Terminations cut off potential feedback loops in the system which, in turn, preclude the possibility of improving on the comprehensiveness and the quality of health data in the system (Piovesan 2017).

A data system with too many breaks and terminations also limits the possibility to improve the relevance of health data for hyperlocal communities. Strong upward flows feeding national-level indicators which are to be used at the regional and local levels for planning and response purposes, lack relevance at the hyperlocal level and more so if local communities are unable to feed into the system data and information directly related to their day-to-day challenges.

Local health facilities and other actors active at the local level in the health system cannot access granular, up to date data on all health facilities via the dhis2. Selected open health data is, however, available on two public web portals, as well as from other government and supranational sources of open health data.

The presence of open health data on national and supranational health portals shifts the focus from the internal flows of health data as described above to the effects of broader, global forces on the flow of data.

There is no coordination between the available open data sources and they are not used by the health facilities, CSOs or other intermediaries serving the local community. This draws attention to the limitations of open data to disrupt data flows determined by the priorities (programmes) of global networks. If data are made openly available but lack value to local communities in the sense that the data is both useful and usable, then open data is unlikely to increase the chances of local community participation in global networks. This finding draws attention to the fact that the digital divide is not solely the product of capacity constraints at the local level but also of the determination of what data is deemed too valuable, what data is collected, who collects and processes it, and how and by whom it is interpreted at supranational and national levels far removed from the needs and interests of local communities.

Conclusion

The findings of this study on the flow of health data in Tanzania show predominantly unidirectional, upwards and outward flows of data to support the suggested systemic cleavage between hyperlocal communities and globally networked actors. Global health indicators determined at the global level, combined with funding for prioritised health areas, serve to shape national data systems and the flow of data within those systems. Open health data, while clearly available in Tanzania, flows between national and supranational actors but does not initiate new flows of health data to the hyperlocal level.

The findings highlight the need for network switchers, possibly at the local level, who are able to connect the global (the space of flows) with the hyperlocal (the space of places); and for network programmers, possibly at the national and supranational levels, who are able to challenge the power of existing global networks as new data flows are activated. Without changes to the current health data system in Tanzania, demand for data at the hyperlocal level will remain poorly defined and thus the flow of data will continue to reflect the needs of those more powerful actors located elsewhere in national networks that are increasingly tied to global networks.

Acknowledgements

The funding provided by the United States President's Emergency Plan for AIDS Relief (PEPFAR) and the Millennium Challenge Corporation (MCC) via their Data Collaborative for Local Impact (DCLI) programme's DataZetu project is acknowledged, as is the support of the International Development Research Centre (IDRC) and Open Data for Development (OD4D) in the presentation of a draft of this chapter at the Open Data Research Symposium in Buenos Aires in September 2018.

REFERENCES

Bates J, Lin Y-W & Goodalde P (2016) Data journeys: Capturing the socio-material constitution of data objects and flows. *Big Data & Society*: 1–12

Bhatia V, Stout S, Baldwin B & Homer D (2016) *Results Data Initiative: Findings from Tanzania*. Washington DC: Development Gateway

Boerma T (2013) Public health information needs in districts. *BMC Health Services Research* 13 (sup.2): 12. http://www.biomedcentral.com/1472-6963/13/S2/S12

Castells M (1998) *End of Millennium. The Information Age: Economy, Society and Culture*. Volume 3. Oxford: Blackwell

Castells M (2009) *Communication Power*. Oxford: Oxford University Press

Castells M (2010) *The Rise of the Network Society. The Information Age: Economy, Society and Culture*. Volume 1 (revised edition). Oxford: Blackwell

Castells M (2017) Afterword 2017. In: J Muller, N Cloete & F van Schalkwyk (eds) *Castells in Africa: Universities and Development*. Cape Town: African Minds. pp. 197–201

Castells M & Himanen P (2014) Models of development in the global information age: Constructing an analytical framework. In: M Castells & P Himanen (eds) *Reconceptualising Development in the Global Information Age*. Oxford: Oxford University Press. pp. 7–25

Custer S, King EM, Atinc TM, Read L & Sethi T (2018) *Toward Data-Driven Education Systems: Insights into using information to measure results and manage change*. Brookings Institution and AidData

Data Zetu (2017a, September) Listening campaign guide. https://docs.google.com/document/d/1lpSr24DrO50bvU1k6qNp5wIPhQ2mTRH0fYOwmNPC7tI/edit

Data Zetu (2017b, 12 October) Pain Points Prioritization Guide: Data Zetu, Kyela. https://docs.google.com/document/d/1GBm5KQVtWzPmqaqdcP3h6CLxDOy_P-5QAr8rHFaxKHw/edit?ts=59e811f3#heading=h.gjdgxs

Dodds L & Wells P (2019) Issues in open data: Data infrastructure. In: T Davies, S Walker, M Rubinstein & F Perini (eds) *The State of Open Data: Histories and horizons*. Cape Town & Ottawa: African Minds and International Development Research Centre. pp. 260–273

ECOSOC (2015) Citizen-based Monitoring of Development Cooperation to Support Implementation of the 2030 Agenda. *2016 Development Cooperation Forum Policy Briefs No. 9*. New York: Development Cooperation Policy Branch, ECOSOC

Harrison TM, Pardo TA & Cook M (2012) Creating open government ecosystems: A research and development agenda. *Future Internet* 4(4): 900–928

Heimstädt M, Saunderson F & Heath T (2014) From toddler to teen: Growth of an open data ecosystem. *JeDem-eJournal of eDemocracy and Open Government* 6(2). DOI: https://doi.org/10.29379/jedem.v6i2.330

Kimunai E (n.d.) Availability of health data. Blog: My One Year of Service – Dodoma, Tanzania. https://myoneyearofservice.wordpress.com/2014/09/16/availability-of-health-data/

Kitchin R & Lauriault TP (2014) Towards critical data studies: Charting and unpacking data assemblages and their work. *The Programmable City Working Paper 2*. http://mural.maynoothuniversity.ie/5683/1/KitchinLauriault_CriticalDataStudies_ProgrammableCity_WorkingPaper2_SSRN-id2474112.pdf

Ministry of Health and Social Welfare, National Aids Control Programme, United Republic of Tanzania (2017, January) *National Guidelines on HIV and Aids Data Management*. Dar es Salaam: Ministry of Health and Social Welfare

Ministry of Health and Social Welfare, United Republic of Tanzania (2008) *Mapping of partners and financial flows in the medicines procurement and supply management system in Tanzania*. Dar es Salaam: Ministry of Health and Social Welfare

Ministry of Health and Social Welfare, United Republic of Tanzania (2009) *Proposal to Strengthen Health Information System (HIS)*. Dar es Salaam: Ministry of Health and Social Welfare

Ministry of Health and Social Welfare, United Republic of Tanzania (2015) *Health Sector Strategic Plan July 2015–June 2020 (HSSP IV)*. Dar es Salaam: Ministry of Health and Social Welfare

Ministry of Health, Community Development, Gender, Elderly and Children (MOHCDGEC), United Republic of Tanzania (2016) *The National Road Map Strategic Plan to Improve Reproductive, Maternal, Newborn, Child & Adolescent Health in Tanzania (2016-2020): One Plan II*. Dar es Salaam: Ministry of Health, Community Development, Gender, Elderly and Children

Ministry of Health, Community Development, Gender, Elderly and Children (MOHCDGEC), United Republic of Tanzania (2017) *The National Health Policy (Sixth Draft Version)*. Dar es Salaam: Ministry of Health, Community Development, Gender, Elderly and Children

Ministry of Health Community Development, Gender, Elderly and Children, United Republic of Tanzania (2018) Tanzania Health Data Collaborative Communique on Commitments to Support One Monitoring and Evaluation Framework for the Health Sector. Dar es Salaam: Health Data Collaborative

Mori AT, Kaale EA, Ngalesoni F, Norheim OF & Robberstad B (2014) The role of evidence in the decision-making process of selecting essential medicines in developing countries: The case of Tanzania. *PLoS ONE* 9(1). DOI 10.1371/journal.pone.0084824

PATH (2017) *Journey to Better Data for Better Health in Tanzania 2017-2023*. Data Use Partnership. https://www.path.org/resources/data-use-partnership-the-journey-to-better-data-for-better-health-in-tanzania/

Piovesan F (2017) Beyond standards and regulations: Obstacles to local open government data initiatives in Italy and France. In: F van Schalkwyk, S Verhulst, G Magalhães, J Pane & J Walker (eds) *The Social Dynamics of Open Data*. Cape Town: African Minds. pp. 35–62

Ravallion M (2016) Are the world's poorest being left behind? *Journal of Economic Growth* 21(2): 139–164

Sato M, Maufi D, Mwingira UJ, Melkizedeck T, Leshabari MT, Ohnishi M & Honda S
(2017) Measuring three aspects of motivation among health workers at primary level
health facilities in rural Tanzania. *PLoS ONE* 12(5). https://doi.org/10.1371/journal.
pone.0176973

Scrollini F (2017) Open your data and will 'they' build it? A case of open data co-
production in health service delivery. In: F van Schalkwyk, S Verhulst, G Magalhães,
J Pane & J Walker (eds) *The Social Dynamics of Open Data*. Cape Town: African Minds.
pp. 139–152

Silaa R & Van Schalkwyk F (2018) Mapping Hyperlocal Health Data Flows: The Case
of Kyela District, Tanzania. Dar es Salaam: Data Zetu. https://drive.google.com/
file/d/1ESwrmdQYj0rjeQ0_Ws4cbKZMLlk6SQ8Q/view

Tunga M & Mushi J (2016) *Towards engaging civil society organisations (CSOs) in the open
data agenda: A case study of selected CSOs in the health sector in Tanzania.* Washington
DC: World Wide Web Foundation

Unicef (2016) *Unicef Annual Report 2016: Tanzania.* Unicef. https://www.unicef.org/
about/annualreport/files/Tanzania_(United_Republic_of)_2016_COAR.pdf

United Nations Economic Commission for Africa (ECA) (2017) *Africa Data Revolution
Report 2016*. Addis Ababa: United Nations Economic Commission for Africa

USAID (2015) *eLMIS Selection Guide.* Arlington, VA: USAID Deliver Project

Van Schalkwyk F & Cañares M (2020) Open data and social inclusion. In: ML Smith &
RK Seward (eds) *Making Open Development Inclusive.* Cambridge, MA: MIT Press

Van Schalkwyk F, Verhulst S & Young A (2017) South Africa's medicine price registry.
In: S Verhulst & A Young (eds) *Open Data in Developing Economies: Toward Building an
Evidence Base on What Works and How.* Cape Town: African Minds. pp. 152–167

Van Schalkwyk F, Willmers M & McNaughton M (2016) Viscous open data: The roles of
intermediaries in an open data ecosystem. *Information Technology for Development* 22
(sup.1): 68–83. DOI 10.1080/02681102.2015.1081868

Verhulst S & Young A (eds) (2017) *Open Data in Developing Economies: Toward Building
an Evidence Base on What Works and How.* Cape Town: African Minds

Verhulst S, Noveck B, Caplan R, Brown K & Paz C (2014) The open data era in health
and social care. NYLS Legal Studies Research Paper No. 2563788. *Social Sciences
Research Network.* https://ssrn.com/abstract=2563788 or http://dx.doi.org/10.2139/
ssrn.2563788

Wickremasinghe D, Hashmi IE, Schellenberg J & Avan BI (2016) District decision-
making for health in low income settings: A systematic literature review. *Health Policy
and Planning* 31: ii12–ii24

World Health Organisation (n.d.) Tanzania Country Profile: Health information,
research, evidence and knowledge. World Health Organisation (WHO). http://www.
aho.afro.who.int/profiles_information/index.php/Tanzania:Health_information,_
research,_evidence_and_knowledge#Data_sources_and_generation

8.

Decentralised open data publishing for the public transport route planning ecosystem

Julián Rojas, Bert Marcelis, Eveline Vlassenroot,
Mathias van Compernolle, Pieter Colpaert & Ruben Verborgh

Open data initiatives have created a revolution in the route planning ecosystem for the public transport sector. The creation of a large amount of route planning services like Google Maps, CityMapper or Navitia, has only been possible thanks to the availability of public transport data as open data. Ever since the disclosure of the London public transport data sources as open data (Hogge 2016) more public transport companies are following their lead around the world. The benefits obtained by disclosing public transport datasets as open data are diverse and influence the different actors present in the route planning ecosystem: public transport organisations in the role of data publishers for instance may increase their revenue streams as new and better information channels attract more travellers (UK Department for Transport et al. 2018). Also, new analysis and improvements to their operations become possible through feedback received from data reusers on areas where they do not collect data by themselves (e.g. crowdsourced data).

For common travellers the benefits are reflected on a more diverse service offer that covers a wider range of functionalities and facilitates ubiquitous access to public transport data through mobile applications. For example, the GoOV[1] application in the Netherlands provides support for anyone who has trouble travelling independently throughout the public transport network, like people with disabilities or seniors. The application relies on public transport open data to guide its users and provides a service to a more specific target group that was not offered before by the public transport operators in the Netherlands. The release of public transport datasets as open data has proven also to be a

---

1   http://www.go-ov.nl/

catalyser for innovation and an economy booster, as revealed by a study on the impact of opening up public transport datasets in London (Deloitte 2017). Over 13 000 registered developers or reusers have contributed to the creation of more than 600 applications that rely on the open data, reaching 42% of London's population and providing innovative commercial and non-commercial customer-face solutions that can tackle social and economic issues. This contributes to the digital economy of the city with an estimate of 500 direct and 230 indirect jobs and an estimated total gross value add from these companies, directly and across the supply chain and wider economy of £14 million per year (Deloitte 2017). Finally, open public transport data represent a valuable source of information for public authorities and NGOs who may use it during decision-making processes (e.g. urban planning) and for independent analysis and studies where public transport is relevant (Share-PSI 2016).

The existence of open data provides a continuum of value. The final parts of the value chain, which involve extracting meaning from data and applying it to address a particular matter, are as important as the earlier parts, which involve data collection, storage and publication (Van Schalkwyk et al. 2017). From a technical perspective in the public transport sector, the way open data is published directly influences the architectural design of route planning applications, which in turn affects the technical decisions that data reusers need to make when using open data.

On one hand, public transport operators may choose to share their data through Remote Procedure Call (RPC) APIs. In the public transport environment, an example of an RPC API is one that receives requests containing a set of parameters (e.g. origin, destination, departure time, etc.) from a remote client, such as a mobile or web route planning application, and uses them to calculate route alternatives over a transport network. Besides routes, RPC APIs could also allow reusers to access information about other related entities (e.g. stops, vehicles, departures, etc.) that can be integrated in their applications. However, with this approach, operators often impose querying limitations to reusers due to the associated computational costs that will increase as the amount of reusers grows. Such limitations go against the idea of open data, the proponents of which advocate for full and unlimited access to data. Furthermore, reusers are not able to influence the route planning algorithms to include new features (e.g. wheelchair accessibility, foldable bikes, shared cars, etc.) as these are perceived as black-boxes from the reusers' perspective.

On the other hand, operators can share their entire datasets using standard formats like General Transit Feed Specification (GTFS)[2] which third parties can integrate and reuse in their applications. Such a data dump approach fosters the creation of centralised data silos, as route planner developers need to process and host the entire dataset of every public transport network over which

2   https://developers.google.com/transit/gtfs/

156

they want to provide their service. Data silos are the result of data integration processes, where it is first necessary to align and reconcile data entity identifiers coming from different data providers, in order to enable route planning queries. For applications that ultimately want to provide a world-wide route planner, this means an immense investment on computational infrastructure.

Considering these approaches and their limitations, the Linked Connections[3] (LC) specification was introduced. LC aims at offering an in-between solution, that is between the RPC API supporting any type of query strategy and the data dump containing all data approach, that allows operators to share data in a cost-efficient way and that is optimised for performing route planning algorithms. By modelling transport networks as a list of vehicle departures and arrivals, sorting them in a timely fashion and publishing them as data fragments, reusers are able to request specific parts of a transport network dataset over which they can calculate a specific route on the fly. LC follows the Linked Data principles[4] by assigning unique identifiers to every element of a public transport network and relying on common semantic vocabularies to provide a description to each of them. This is intended to increase the interoperability of public transport datasets which reduces at the same time the adoption costs of data for open data reusers.

The approach of fragmenting datasets was taken from the linked data fragments concept (Verborgh et al. 2014) which allows for the definition of specific types of fragments of linked data datasets that can be generated with minimal effort by servers, while still enabling efficient client-side querying. This constitutes a decentralised solution as reusers can now directly request specific data fragments from different public transport operators that are distributed on the web and execute route planning algorithms just in time on the client side, therefore reducing data hosting and integration costs. Furthermore, LC allows clients to cache fetched data fragments in memory, enabling offline execution of new queries, which is not possible on RPC based solutions.

The LC framework also provides a solution for route planning that supports *privacy-by-design*. Since the route planning calculations can be performed on the client side, the users are not required to share the details of their queries with third party servers. Previous research (see Colpaert et al. 2017) has proven that from a scalability and cost-efficiency perspective for hosting the data, Linked Connections outperforms traditional RPC based route planning approaches. This is an advantage for data publishers as they are able to provide data to more reusers with lower operational costs. However, it is still unclear how an approach such as LC will impact other actors in the route planning ecosystem for public transport (e.g. reusers, common travellers).

Roy Fielding (2000) introduced Representational State Transfer (REST) in his PhD thesis while standardising the web's HTTP/1.1 protocol. REST is a

---

3   http://linkedconnections.org/
4   https://www.w3.org/DesignIssues/LinkedData.html

set of architectural constraints for large data architectures. Each constraint one decides to follow in a web architecture will return benefits such as scalability of the server, visibility, cost-efficiency, reliability and – also – *the user-perceived performance*. For instance, following the caching constraint on both client and server sides results in (i) more scalable servers by reducing the amount of requests that servers need to process, as the number of clients increases; and (ii) improved user-perceived performance by allowing clients to keep and reuse relevant data from memory instead of requesting it from the server every time it is needed, which is significantly slower. In this chapter we particularly zoom in on the *user-perceived performance* property and study it within the context of open data for route planning purposes.

The main research question we address in this work is: what is the impact on the actors that belong to the route planning ecosystem for public transport, of implementing an open data publishing approach as the Linked Connections framework? In this work we present and discuss an analysis of such effects. We also present a study that evaluated the technical performance of an LC based application that executes its route planning algorithm on the client side compared to a traditional application that relies on a RPC route planning API running on the server-side to determine what kind of considerations developers and data reusers must take into account when working with this approach.

Furthermore, we present the results of a user perceived performance study where 17 different regular public transport travellers tested both applications for different use cases and selected one or the other as their preferred choice based on perceived performance and provided features, in order to determine the effects of the LC approach on common public transport riders and their perception of it. For this we developed an isomorphic Android application that implemented both approaches and provided users with the same interface in both cases. In the next session we describe the open data route planning ecosystem for the public transport sector and the different actors that comprise it. Then we present a description of the methodology followed during the performed studies. The results obtained during the evaluations are presented afterwards. Finally, a discussion of the main findings is presented along with the correspondent conclusions.

The route planning ecosystem

The use of the ecosystem analogy in relation to business practices has become notably strong. Related literature defines digital ecosystems as cyclical, sustainable, demand-driven environments orientated around agents of a different nature who are mutually interdependent (Heimstädt et al. 2014). Scholars in information intensive environments have used the term to focus on the multiple and varying interrelationships between providers, users, data, infrastructure and institutions (Harrison et al. 2012). For open data route planning on the public transport sector we devise an ecosystem as shown in Figure 1, where the different actors that benefit from and support open data, are represented.

Figure 1. Open data route planning ecosystem

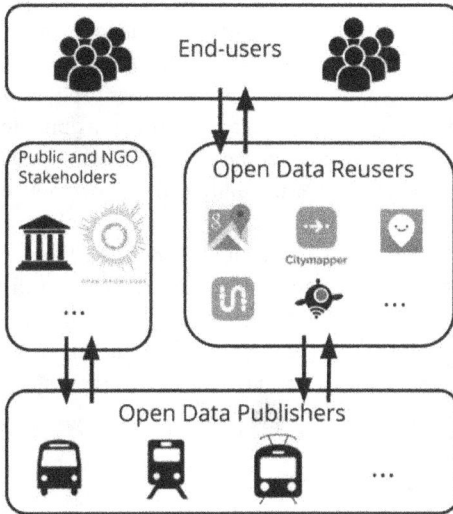

The first type of actors that can be identified in Figure 1 are the *open data publishers*. On a route planning ecosystem these correspond to the public transport companies which operate on a transport network infrastructure (e.g. bus, train, tram, metro, etc.) and produce related data (e.g. timetables, list of stations, live updates, etc.). They publish the data as open data in machine readable formats that allow its adoption and reuse by third parties. The GTFS specification, as the *de facto* standard, is commonly the selected format to publish and share the data.

The *open data reusers* reference every company and/or organisation that consumes and integrates open data for solving route planning queries over one or more public transport networks. Here we can find companies such as Google (Maps), CityMapper, Moovit, GoOV and many others of the sort, that collect and integrate public transport-related open data to offer route planning services on top of it. Public authorities also reuse public transport open data to offer route planning services as a mechanism to improve the mobility conditions of their regions and cities (Rode et al. 2015; Ahlers et al. 2018). Public authorities in cities such as Portland[5] (US), Antwerp[6] (Belgium) and London[7] (UK) can be considered as examples of open data reusers.

---

5   https://trimet.org/#/planner
6   https://www.slimnaarantwerpen.be/en/home
7   https://tfl.gov.uk/plan-a-journey/

The *end-users* on a route planning ecosystem are the public transport travellers. These actors use the services offered by open data reusers through web or mobile applications to navigate the public transport networks and satisfy their mobility needs. Lastly, the *Public and NGO Stakeholders* normally do not have direct contact with the data flow through the route planning ecosystem (except when acting as data reusers) but can influence how the data is shared among its actors. Public authorities, for instance, define the legal framework and constraints of data sharing processes, and other types of organisations such as research institutes and non-profit agencies can contribute to the definition of standards and procedures that impact the way open data is shared and consumed.

This ecosystem constitutes the analytical framework of our work. Its definition has been made based on an empirical mapping of real-world route planning-related scenarios, by observing the relationships and interactions of the actively involved organisations. Determining how the different actors that comprise it are affected by the implementation of a decentralised open data publishing strategy represents a contribution to the open data community on the public transport sector by shedding light on the merits and also on the open challenges of the aforementioned approach.

Methodology

In this section we describe the methods used for assessing the impact of implementing a decentralised open data publishing strategy within the route planning ecosystem for public transport and its actors. Having identified the different actors that play a role in the ecosystem, in this work we focused specifically on assessing the impact on open data reusers and end-users.

Open data reusers

For conducting the impact evaluation on open data reuser tests we developed an isomorphic Android application[8] that implemented an LC based route planner and a client for an RPC based API hosted on a remote server (see Figure 2). The evaluation consisted of a series of performance tests that were conducted for both the LC and the RPC API based implementations. We used the same algorithm and datasets in both cases to allow a fair comparison. The chosen algorithm was the Connection Scan Algorithm (CSA) which was designed to operate over similar data structures as the one defined by the LC specification, making it a perfect fit. Moreover, previous research (see Dibbelt et al. 2018) has proven the CSA algorithm to be more efficient for public transport route planning than traditional route planning solutions based on variants of Dijkstra's algorithm (Dijkstra 1959).

---

8   Available at https://github.com/Bertware/masterthesis-LC-LC2Irail-android-client/releases/
     tag/RESEARCH40

Therefore, we propose the following hypothesis:

> *(H1): LC based implementations perform better with regard to response time when compared to traditional RPC API based implementations.*

For the performance tests we defined a set of use cases to be tested on both implementations. These consisted of queries that requested routes (going from A to B at a given departure time), liveboards (the list of scheduled arrivals and departures at a given stop) and vehicles (the sequence of stops and scheduled times for a given vehicle). The dataset used for the evaluation is the public train transport network of Belgium operated by the NMBS company.[9] The set of queries used to test the different use cases were taken from real world requests made to the servers of the iRail API.[10] Both implementations were tested on two different smartphones with different hardware capabilities: the HTC One (Android 5.0) and the HTC 10 (Android 8.0) smartphones.

Figure 2. HyperRail: Isomorphic route planning mobile application

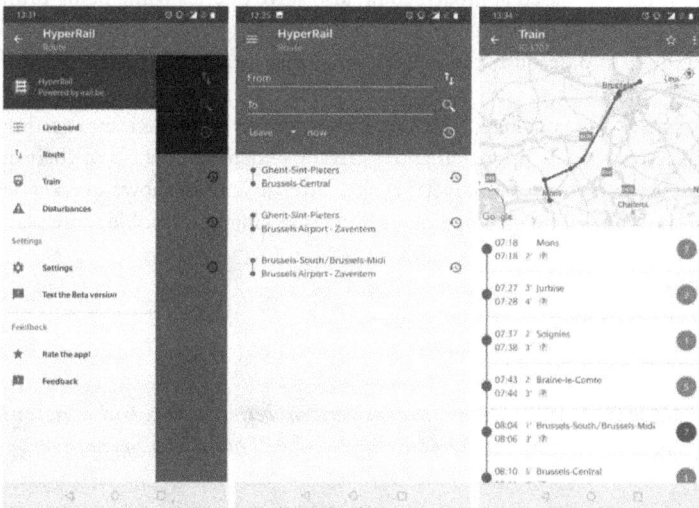

Open data end-users

The LC approach publishes raw public transport schedules as data fragments allowing route planning application clients to request only the required data to solve specific queries. Clients can cache these data fragments in memory

---

9   NMBS Linked Connections available at https://graph.irail.be/sncb/connections
10 The iRail project: https://hello.irail.be

and reuse them to solve future queries, accelerating the process of solving route planning queries. Considering this, we propose the following hypothesis:

(H2): The LC based route planning application has a higher user perceived performance when compared to traditional RPC API based implementations.

The evaluation of the impact on end-users was carried out through a user perceived performance test (n=17) and a questionnaire (n=65). For the user perceived performance test we used the same isomorphic Android application used for the performance evaluation for open data reusers, as well as the same defined use cases and dataset. Each user was asked to execute a set of queries for each use case using both the RPC API based and the LC based approaches. Then for each use case, every user was asked to provide their opinion on which alternative they perceived to perform better. Furthermore, once the users completed testing each use case, they were asked to activate the airplane mode of the smartphone and to run the queries again for both approaches. This was done in order to show the users that the LC based approach could also solve queries while being offline.

Additionally, all the users had to decide on which approach they preferred the most, taking into account the performance they perceived during the different use cases and also the additional features, such as offline querying and privacy safeguarding. To conclude, the users received an explanation regarding the capabilities of the LC approach about privacy (where the details of their queries were not being sent to a remote server), speed (results can be shown quicker because LC can load and reuse information), offline querying and flexible route planning.

Findings

Open data reusers

(H1): LC based implementations perform better with regard to response time when compared to traditional RPC API based implementations.

Figure 3 shows the results of the evaluation performed when querying for routes. The median response times are depicted for both approaches running on the HTC 10 and HTC One smartphones. In the HTC One, the LC approach performs 29% faster (~1s) than its counterpart in the HTC 10. However, for the HTC One the performance of LC deteriorates being 57% slower (~2s) than the RPC API. This behaviour can be explained by the fact that the HTC One smartphone has inferior hardware capabilities which impacts the execution of the route planning algorithm on the device. Also, as expected, the performance of the RPC approach is consistently similar on both devices as the algorithm is executed on the server-side.

Figure 3. Number of results in function of loading time (routes) – LC vs RPC API

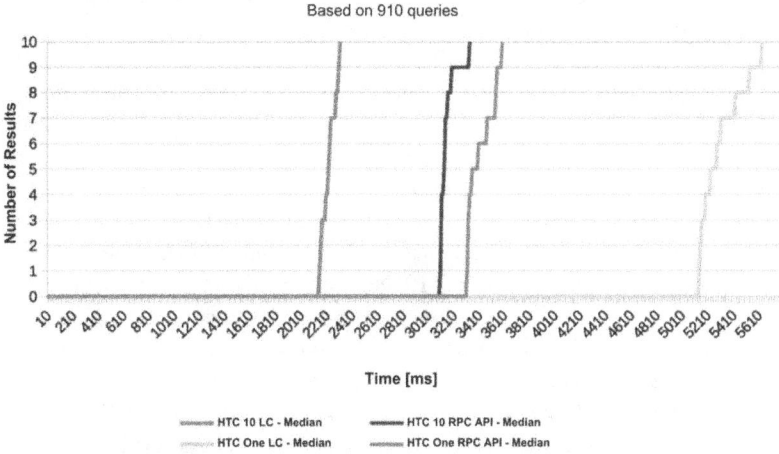

Based on 910 queries

Figure 4 shows the results of the evaluation performed for *liveboards* queries. The median response times are depicted for both approaches running on the HTC One and HTC 10 smartphones. The incremental results make clear that there are differences between both architectures. Linked Connections is faster than the RPC API based approach on both devices for the first few results. However, when ten results are needed (about the number of results which fit on a large screen) the RPC API is faster on both devices. It is clear that the RPC API performs similarly on both devices, but the LC client does not. There is a gap between the time needed by the client-side LC implementation on both devices, which grows with the number of results needed.

Figure 4. Number of results in function of response time (liveboards) – LC vs. RPC API

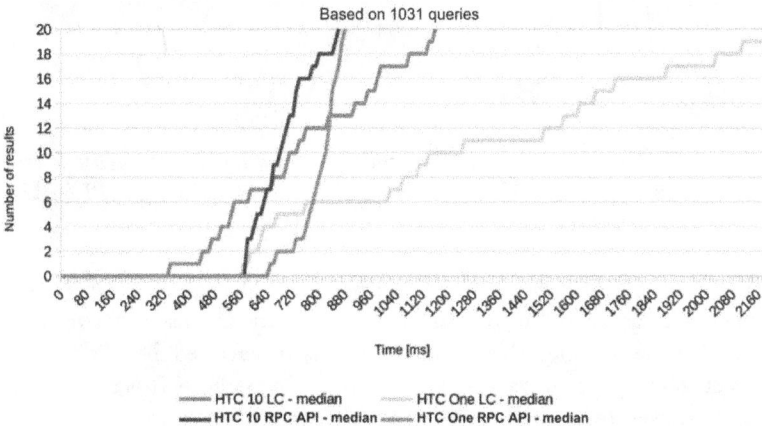

Based on 1031 queries

163

Figure 5 shows the distribution of response times for *vehicle* queries for both approaches and in the devices used for the tests. Every vehicle trip is considered to be one atomic result; therefore, incremental results are not supported for this data type. When looking at the distribution of the response times it becomes clear that vehicle data take a longer time to load using the LC implementation. The information about a single vehicle typically spans around 3 or 4 hours, which translates into a larger amount of data fragments that need to be retrieved and processed. The RPC API based approach, which has quicker access to the data, has an advantage here. It also performs consistently between devices, whereas the LC implementation needs two times as much time on the HTC One, compared to LC implementation on the HTC 10. Not only the data type and device affect the performance, but the exact query is of importance too. Calm stations, long routes, or vehicles with a long trip take longer to load compared to busy stations, short routes or vehicles with a short trip. The time to load a number of results is directly related to the timespan in which the results can be found. When a larger timespan needs to be evaluated, the results will take longer to load.

Figure 5. Technical performance distribution of vehicle queries on LC vs RPC API

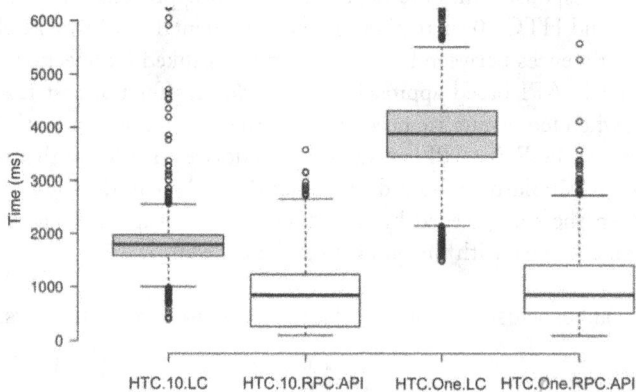

Open data end-users

> *(H2): The LC based route planning application has a higher user perceived performance when compared to traditional RPC API based implementations.*

Figure 6 depicts a summary of the user perceived performance tests for every defined use case and the overall choice made by the users between both approaches. Results show that the majority of users perceived the RPC approach as faster in every tested use case (*liveboards* – 47%, *routes* – 76% and *vehicles* – 65%). However, the overall choice shows that 59% of the users picked the LC approach as the preferred choice. For most of them, this final choice was made mainly due to capacity of queries executed offline. Another reason for users to have given preference to the LC approach is privacy, where 85% expressed that they would be bothered if their location would be sent over the internet and 77% would be bothered if their journey itineraries would be known by third parties, which is the case for most route planning applications nowadays.

Figure 6. User perceived performance results for *liveboards, routes, vehicle* queries and overall choice between LC and RPC API

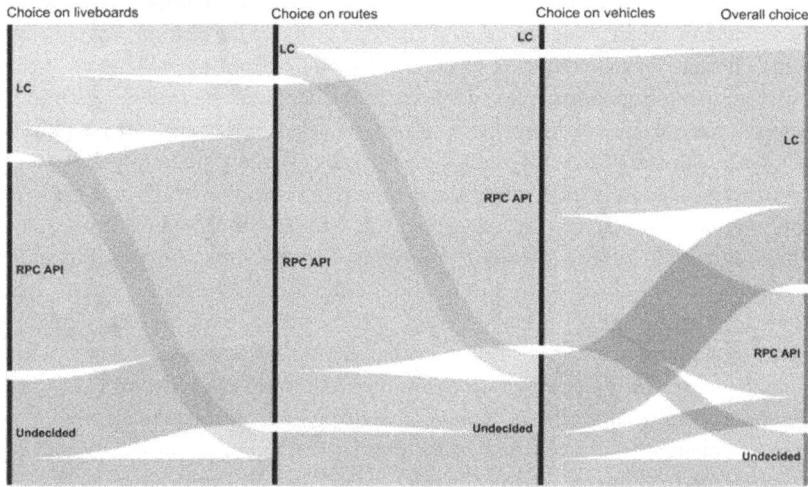

Discussion and conclusions

In order to determine the impact of implementing a decentralised publishing strategy of public transport open data for the route planning ecosystem, such as the Linked Connections framework, we conducted a series of evaluations that focused on the *open data reuser* and the *open data end-user* actors. However, even though we did not assess the impact of the LC approach on *open data publishers*, we can refer to previous work where the cost-efficiency of implementing the LC approach was measured (see Colpaert et al. 2017). Results showed that for data publishers, following the LC approach meant lower infrastructure associated costs as they can support a larger number of requests with less powerful servers, thus having better scalability. This has a positive and important effect for open data publishers, as one of the main goals of open data is to maximise data reuse and with this approach they can now support a larger number of clients with a lower investment.

Moreover, unrestricted access to data, which is one of the main challenges of open data in the route planning ecosystem, is also tackled by the LC approach.

Most traditional approaches use RPC API based architectures to expose route planning data and often require the imposition of access restrictions (e.g. in terms of number or requests per day) to their users to prevent overloading their servers. But with the higher efficiency achieved by the LC approach, open data publishers can give unrestricted access to the data mainly because the data becomes cacheable and the processing load of calculating routes is now moved on to the client side. But unrestricted access to data can be also interpreted from a query flexibility perspective. With an RPC API based approach, *open data reusers* are limited by the type of queries that the API has been built to support and cannot influence the type of data they obtain from each query. For route planning this means that *open data reusers* can request, for example, data about route alternatives to go from A to B from the route planning RPC API of the buses and trams operator, but cannot ask to include bike or car sharing options into the route calculation process. For an *open data reuser* to support new kinds of queries, this traditionally means creating a new route planning API from scratch and manually integrating the different datasets they want to include in their queries. The LC approach leverages this issue by simply publishing the raw data fragmented following a strategy optimised for route planning purposes. In this way, *open data reusers* can directly access the specific parts of a dataset that they need and combine them with any other external data sources, allowing them to support new types of queries. For example, an *open data reuser* could directly reuse the LC dataset from the bus and tram operator of a city and combine them with available bike or car sharing open datasets to support new types of queries and render new route alternatives, without being restricted by precalculated routes offered by their RPC API or the overhead of having to integrate the complete buses and trams dataset first.

We also did not focus on the *public and NGO stakeholder* actors. As mentioned before, these actors contribute to the ecosystem by providing the legal framework and the definition of mechanisms and standards through which the route planning ecosystem is supported. Therefore it can be argued that since they do not take a direct active role in the open data flow that takes place inside the route planning ecosystem, there would be no significant impact to this institution when implementing a decentralised open data publishing strategy.

When looking at the results obtained in the evaluation for *open data reusers*, we can observe that a route planning application implementing the LC specification and processing queries on the client side, performs acceptably well compared to its RPC API based counterpart, even obtaining a better performance for *routes* queries. We take into account that we only measured the app performance based on response time. Other performance benchmark methodologies could be used to get a more detailed insight into the performance, for example, bandwidth usage or battery consumption on end-user devices. Ensuring high performance of the applications is a main concern for reusers who seek to provide a high quality of service to their users and the results obtained during these evaluations show that the LC approach provides a feasible alternative for the route planning ecosystem. However, there are still some types of queries where the LC approach does not perform as well as its counterpart, like for *vehicle* queries. Technically, this is due to *vehicle* queries needing access to data from larger timespans than other types of queries, which requires LC clients to request and process a higher amount of data fragments. This lays out a gap in the design of the LC specification that needs to be addressed by developers when implementing the specification, and since the LC framework is available as open source, reusers can keep optimising their implementations according to their needs. But without a doubt the greatest benefit that *open data reusers* get from following the LC framework is the full flexibility of data with low adoption costs. With LC, reusers can access the raw data from one or more public transport networks, which is not possible on traditional RPC API based approaches where the data flexibility is constrained by API implementations. Moreover, reusers can implement their own algorithms, integrating any kind of external data they could need to offer a specific service. Also, by publishing data as fragments and following the Linked Data principles, reusers can access the specific portions of data they need to solve any given query while using a unified model supported by semantic vocabularies. This lowers the data adoption costs for community reusers who do not need to incur data hosting and integration costs, allowing them to focus on the development of their core service (i.e. route planning algorithm). Considering this, it is possible to argue that a decentralised open data publishing strategy, such as the LC framework, may contribute to innovation and thus, the economic growth of the route planning ecosystem.

On the other hand, the results of the user perceived evaluation carried out for *open data end-users* shed light on the fact that the majority of the users in this study will value the additional features (such as offline querying and the safeguarding of their privacy) more than the performance. It is important to note that the empirical results reported in the research are subject to several limitations. First, there is the low number of participants. Only 17 respondents participated in the user perceived performance test and 81 in the questionnaire. Therefore, it is difficult to identify significant relationships in the data.

The second limitation concerns the internal validity of the user perceived performance test with regards to the offline and privacy features. In this case, the work of Nissenbaum (2011) is worth mentioning, as users are willing to give up privacy depending on the benefit they receive from a service. This means LC takes privacy more into account as the data, such as location or travelling preferences, are processed at the client side without the need of sending them to remote or third-party servers. Next to this we also note that the hardware capabilities of the end-user device have a major impact on the performance of client-side query evaluation. A less powerful device can reduce the user perceived performance of an LC client, as evidenced by most of the users selecting the RPC based approach as their preferred choice for performance.

This is an important aspect to be considered as not all users may have access to powerful devices, therefore it is an open issue for the LC framework to improve the performance of route planning use cases on devices of lower capabilities.

However, being capable of executing queries offline is a feature that the majority of the users regarded as more important than a better performance due to the fact that when travelling throughout public transport networks, mobile data connections are often lost. This can be noticed, for example, in rural zones with poor coverage or in subway tunnels which render RPC based route planning applications useless. In that case, an LC based client may already have pre-fetched the data or can use the data from a previous look up to keep answering queries. Pre-fetching is hardly possible in an RPC-style API, as this would require a request per every possible query. Also, being allowed to be in control of their own personal data (e.g. location and itineraries) is an important factor for end-users. With the recent breaking out of scandals about how personal data being collected through social media was being used to influence election results all over the world, users have become more aware of the importance of their data privacy.

Therefore, LC based applications provide them with a good alternative that takes this matter into account and protects the very sensitive data that is required in the route planning ecosystem. From a general perspective we could state that the impact for open data end-users of a decentralised approach would be reflected through a bigger, more varied and more personalised offer of services for the route planning ecosystem.

This work provides insights and an initial assessment of the potential effects of implementing a decentralised open data publishing strategy in the public transport route planning ecosystem. We have been able to observe that even though it still requires further work to improve some identified shortcomings, the potential benefits of such an approach are aligned with the ideals of open data of fostering innovation, boosting economic growth and providing solutions for more specific necessities (e.g. public transport accessibility for people with disabilities). Determining what key aspects end-users value the most when choosing an application and also which factors limit the performance of decentralised approaches are fundamental steps towards building a richer and sustainable route planning ecosystem that increases innovation and adoption of open data in the public transport sector.

However, it still remains as a research interest to determine how the decentralisation of open data publishing can be applied to other sectors. Also, how the actors in the ecosystem behave towards each other and how LC affects their current organisational and business models. At first sight, Linked Data and semantic technologies could provide the means to increase interoperability of datasets but further effort in creating comprehensive and common domain ontologies is still needed. Furthermore, exploring different strategies for fragmenting datasets that suit the needs of other (policy) domains and keep open data adoption costs low, is also an interesting research direction.

REFERENCES

Ahlers D, Akerkar R, Krogstie J, Opdahl AL, Tessem B & Zhang W (2018) Harnessing mobility data in cities: A case study from the Bergen region. *NOKOBIT* 1(26). http://hdl.handle.net/11250/2579413

Colpaert P, Verborgh R & Mannens E (2017) Public transit route planning through lightweight linked data interfaces. Paper presented at the 17th International Conference on Web Engineering, Lecture Notes in Computer Science: 403–411. https://doi.org/10.1007/978-3-319-60131-1_26

Deloitte (2017) Assessing the value of TfL's open data and digital partnerships. http://content.tfl.gov.uk/deloitte-report-tfl-open-data.pdf

Dibbelt J, Pajor T, Strasser B & Wagner D (2018) Connection scan algorithm. *The Journal of Experimental Algorithmics* 23: 1–56. http://doi.acm.org/10.1145/3274661

Dijkstra EW (1959) A note on two problems in connexion with graphs. *Numerische Mathematik* 1: 269–271. http://dx.doi.org/10.1007/BF01386390

Fielding R (2000) Architectural styles and the design of network-based software architectures. Doctoral dissertation, Information and Computer Science, University of California

Harrison T, Pardo T & Cook M (2012) Creating open government ecosystems: A research and development agenda. *Future Internet* 4(4): 900–928. https://www.mdpi.com/1999-5903/4/4/900

Heimstädt M, Saunderson F & Heath T (2014) Conceptualizing open data ecosystems: A timeline analysis of open data development in the UK. Paper presented at the International Conference for E-Democracy and Open Government. https://project.opendatamonitor.eu/wp-content/uploads/dissemination/OpenDataMonitor_Publication_Conceptualizing-Open-Data-Ecosystems.pdf

Hogge B (2016) *Open Data's Impact: Transport for London Get Set, Go!* Govlab & Omidyar Network. http://odimpact.org/files/case-studies-transport-for-london.pdf

Nissenbaum H (2011) A contextual approach to privacy online. *Daedalus* 140(4): 32–48. http://amacad.org/sites/default/files/academy/multimedia/pdfs/publications/daedalus/11_fall_nissenbaum.pdf

Rode P, Hoffmann C, Kandt J, Graff A & Smith D (2015) *Towards new urban mobility: The case of London and Berlin.* London: London School of Economics and Political Science. http://eprints.lse.ac.uk/id/eprint/68875

Share-PSI 2.0 (2016) Best Practice: Open up public transport data. https://www.w3.org/2013/share-psi/bp/ptd/

Van Schalkwyk F, Verhulst SG, Magalhães G, Pane J & Walker J (eds) (2017) *The Social Dynamics of Open Data.* Cape Town: African Minds. http://www.africanminds.co.za/wp-content/uploads/2017/06/9781928331568_txt.pdf

Verborgh R, Hartig O, De Meester B, Haesendonck G, De Vocht L, Vander Sande M, et al. (2014) Querying datasets on the web with high availability. Paper presented at the 13th International Semantic Web Conference, Trentino, 19–23 October

UK Department for Transport, Deloitte & Open Data Institute (2018) Bus Open Data – Collaboration to put the passenger first. https://assets.publishing.service.gov.uk/government/uploads/system/uploads/attachment_data/file/722576/bus-open-data-case-for-change.pdf

9.

Building a framework for the analysis of factors to creation and growth of an open data ecosystem

Edson Carlos Germano, Nicolau Reinhard & Violeta Sun

Public access to large and complex government databases is expected to contribute to the improvement of citizens' lives. However, there are several obstacles to obtaining these expected benefits from open data. On the side of the data producers, these barriers may include problems such as the effort and cost required to convert from closed to open data, and also providing them in a context required by users to ensure their acceptance and usefulness. Other barriers may be poor data quality, lack of legal and political support for the initiative, lack of technical capacity to implement and maintain open data practices and resistance to openness by the databases' administrators (Janssen 2011; Janssen et al. 2012; Magalhães et al. 2013). On the users' side, barriers include problems such as lack of access, low levels of data understanding, lack of human, social and financial resources to effectively use the data, such as combining and integrating different databases to create value for citizens. (Cañares 2014; Gurstein 2011; Magalhães et al. 2013).

Open data intermediaries have been playing a crucial role in removing some of these barriers and thus unlock the potential of open data, by interpreting and integrating open, simple, or complex databases to produce results compatible with users' needs. According to Van Schalkwyk et al. (2014), the presence of intermediaries in the open data ecosystem stimulates the flow of open data among the various ecosystem actors. The publication of databases in open format as well as the intermediaries' effectiveness in generating products and services for end-users have also been improved by more recently developed technologies, such as tools for big data and machine learning. According to the World Bank Report (2016), the internet and digital technologies have spread much more rapidly in developing countries than previous technological innovations. For the generation of digital natives, the natural appropriation of these technological advances reduces the barriers to using open data.

These digital technologies, such as ICT, are incorporated in global, national and local socio-economic contexts that are defined by Diga and May (2016) as ICT ecosystems, encompassing policies, strategies, processes, information, technologies, applications and stakeholders, that together constitute the technological environment for countries, governments or companies. An important aspect of an ICT ecosystem is that it includes people: the various individuals who create, buy, sell, regulate, manage and use the technology.

An ecosystem where open government data (OGD) are the main resources for the establishment of relations between the actors acting from the production of the data in open format to its final consumption as products or services to users will be denominated an OGD ecosystem. This characterisation of the OGD context as an ecosystem allows the representation of it as a broad conceptual framework that considers actors, resource flows, socio-economic, political and spatial dynamics, among other system variables. Its structure and functioning could then be analysed through the study of the existence and intensity of the interactions among the ecosystem's multiple actors. Interactions influenced by systems of governance, citizenship, communication, knowledge, and innovation affecting the OGD ecosystem, whose actors are the producers, intermediaries, and users, and interactions are represented by the flow of information, services, products, and financial resources.

Research problem and objectives

Roberts (2014) points out that citizens are increasingly dependent on third parties, called 'trusted intermediaries', who ensure that transparency policies are maintained and who assist users in accessing the information available in transparency policies, making them accessible and easy to interpret. Roberts further adds the following concern:

> [...] our dependence on intermediaries will increase, and this will raise the difficult question of whether such groups can acquire the resources needed to do the job of intermediation properly. (2014: 10)

In order to understand the different roles played by intermediaries in an OGD ecosystem and their characteristics, the present work proposes a framework considering not only the actors and their relationships. Also under consideration will be variables such as ecosystem governance, existing policies and actors' benefits perception in order to describe the actors, their roles and the structure of the OGD ecosystem, while also identifying variables that can influence this ecosystem and its economic and social sustainability.

Theoretical foundation

The OGD ecosystem analysis framework proposal will be based on the concept of intermediaries and their performance, the use of the ecosystem concept for ICT environments, studies on the OGD value chain, their actors and how value is generated along the chain, and principles for the ecosystem's governance.

Data intermediates

The concept of intermediaries in the field of ICT emerged in the 1980s as a process of intermediation and not as a description of a group of actors, people or organisations that play an intermediary role (Mittilä 2006). The performance of intermediaries was considered by researchers as critical in the production, launching, scalability, and popularisation of innovations through the transmission of information between suppliers. Therefore, intermediaries were seen as liaison organisations, information agents or support organisations positioned between two or more actors (Van Schalkwyk et al. 2016).

According to Davies (2014):

> Intermediaries are vital to both the supply and the use of open data [...] translating data into information, knowledge, and action [...], many citizens' trust in local institutions plays a significant role in their choice to use these [...].

In order to understand the roles played by intermediaries and to justify their existence, Sein and Furuholt (2012) present a set of generalised interpretations of intermediaries: intermediaries 'help users access information that is publicly available by locating these sources', 'integrating various sources on a specific topic, structuring these findings into a form understandable by interested users and disseminating it to them'.

Janssen and Zuiderwijk (2014) in their study of what they described as 'infomediary business models' also considered middlemen as value adding actors positioned between data providers and data users. The authors also highlighted the fact that intermediaries are vital in systems that become increasingly complex, resulting in higher levels of interdependence among multiple agents, with the intensification of individuals' specialisation.

Sein and Furuholt (2012) point out that in the context of e-government and governance, intermediaries perform functions such as: 'diffusion of services (Al-Sobhi et al. 2010, in Sein & Furuholt 2012), reducing corruption (Bhatnagar 2003, in Sein & Furuholt 2012), moderating discussion on democracy (Edwards 2002, in Sein & Furuholt 2012) and providing e-government services of various types (Bailey 2009; Gorla 2009, in Sein & Furuholt 2012)'. These studies mention the 'offline' service offered by intermediaries. (Sein & Furuholt 2012).

Researchers, media, civil society organisations, among others, have a tradition of using government data as a source for their activities which can be seen as pioneering initiatives for open data actions. A study of the practices of organisations using government data has identified the potential of such intermediary organisations to enrich the publication and availability of open data in a precarious ecosystem of suppliers or producers of government open data (Chattapadhyay 2014).

Fransman (2007), based on the work of evolutionary economist Joseph Schumpeter, described the ICT business environment as a sector ecosystem within a larger ecosystem with socio-economic characteristics, with a focus on identifying what drives complex ecosystems with varying degrees of interdependence. In his work, Fransman identified the actors that interact dynamically in the ICT ecosystem, including companies, non-profits, public organisations, consumers and intermediaries connected by exchange relationships, as well as by the set of rules, values, and norms in which the actors are inserted.

According to Fransman (2007), the main characteristic of an ICT ecosystem is its being driven by innovation which is defined as the insertion of new knowledge into the ecosystem. Firms compete and cooperate symbiotically, and the interaction between firms and consumers (that is, between knowledge producers and knowledge consumers) generates new knowledge leading to innovation in the ecosystem. This quest for innovation is that it would keep the ICT ecosystem dynamic and evolving. In developing countries, middlemen play an important role in the innovation environment by filling gaps in the national innovation system (Intarakumnerd et al. 2010, in Intarakumnerd & Chaoroenporn 2013).

Although the aforementioned studies analysed the role of the intermediaries in their sectors of activity, only a few analysed the role of the intermediaries in the context of open data. Van Schalkwyk et al. (2014), in a study on the use of open data in the governance of South African public universities, suggest that an intermediary in this data ecosystem depends on personal connections (or social capital) to enable the flow of data to potential users from government databases. According to the authors, open data intermediaries play several important roles in the ecosystem, such as: (i) increasing the accessibility and utility of data; (ii) assuming the role of a 'keystone species' in a data ecosystem; and (iii) democratising the effects and use of open data (Van Schalkwyk et al. 2014). The authors conclude that, despite the low availability of government data, the open data ecosystem in the governance of public universities has evolved because open data intermediaries have facilitated the data flow from government to potential users.

Johnson (2014) draws attention to the concept of 'disciplinary power' and the potential to locate existing injustices in an open data ecosystem. For Johnson, open data enhance the capacity of disciplinary systems that observe and evaluate the compliance of institutions and individuals with norms that become fundamental values and assumptions of the institutional system, regardless of whether the data reflect the peculiar circumstances of these institutions and individuals. Institutions and individuals who deviate from these norms are marginalised in political debates, since observers evaluate all institutions according to the norm (since there is only data on the actions that reflect the norm) as institutions internalise the norms and guide their actions toward them. Since standards reflect society's power structure where they were created, open data would be reiterating the standards of fairness and injustice.

Studies have sought to create a typology of open data intermediaries that act on specific open data ecosystems. Magalhães et al. (2013) propose a typology of open data intermediaries consisting of three basic types: (i) civic startups, (ii) open data services and, (iii) infomediaries. And they identified three types of business models in companies that use open data as resources for products and services marketed to potential users: enablers, integrators, and facilitators.

Janssen and Zuiderwijk (2014) define infomediaries as the intermediaries that position themselves between data producers and potential users and classified them into six types according to the adopted business model: defined information services, interactive services, aggregators of information, information combiners, open data repositories, and service platforms.

Deloitte Analytics (2012), in collaboration with the United Kingdom Open Data Institute, listed the following types of information intermediaries: suppliers, aggregators, developers, enrichers, and enablers, distributed between two categories of organisations: application developers and businesses.

Using Bourdieu's concepts of social interactions (including the concepts of situations, habitus, and capital), Van Schalkwyk et al. (2016) analysed how open data intermediaries promote the flow of information between data producers and potential users. The study found that the open data value chain may require multiple intermediaries and multiple forms of capital (economic, cultural, social, technical, or symbolic) enabling the connections between data producers and potential users.

The effectiveness of intermediaries in this ecosystem could be attributed to their proximity to data producers or users, and this closeness could be expressed by the type of capital possessed by the intermediary. However, Van Schalkwyk et al. (2016) point out that individually, no intermediary would have all the available capital needed to effectively connect to all agents that exert power in the ecosystem, therefore multiple intermediaries with complementary capital configurations would more adequately represent the various connections between agents that exert the power in the open data ecosystem.

ICT ecosystems

Fransman (2010), describes the components of a socio-economic ecosystem for the ICT context. The author identifies the dynamic interaction between the organisms (companies, non-businesses, intermediaries, and consumers) that compose the ecosystem by grouping them into four layers within an ecosystem that is driven by innovation (e.g. the insertion of new knowledge into the ecosystem):

- network element providers (producing items such as computers, operating systems, mobile phones, telecommunications equipment);
- network operators (creating and operating telecommunications networks, cable TV, and transmission networks);
- content and applications providers, applications, services, innovation, search, navigation and middleware platforms; and
- final consumers.

The layers composed of key groups of the ICT ecosystem are sequentially structured and interdependent. Each layer depends on its adjacent layer (or layers). For the system to function as a whole, each layer must play its functional role, influencing or being influenced by other layers. The division between the layers, however, is not unique as there may exist organisations that act on more than one layer. This ecosystem does not only include the organisations that make up the layers, but also financial institutions, regulatory agencies, standardisation forums, universities, research institutes and other entities (Fransman 2007).

According to the evolutionary theory proposed by Schumpeter, the emergence of new products, new processes, new forms of organisation and new markets is the result of innovation in the system and, in his words, are the main drivers of the evolution of the system itself (Schumpeter 1943, in Fransman 2007).

According to Fransman (2010), these innovations arise essentially from six symbiotic relationships that occur within the ICT ecosystem. The author uses the biology concept of symbiosis to demonstrate that the individuals who make up the ecosystem coexist in constant interaction and the result of these interactions may or may not be beneficial to the parties involved.

These interactions described by Fransman refer to the relationships between the four key stakeholder groups present in the ICT ecosystem: network element providers, network operators, content and application providers, and consumers. According to the author, each of the six relationships can be analysed in four parallel dimensions: (i) financial, through the purchase and sale of products or services, resulting in a financial flow; (ii) material, through the entry and exit of products, resulting in a flow of materials, physical or virtual; (iii) informational, through the exchange of information resulting in a flow of information, and; (iv) innovative, through the creation and diffusion of the ecosystem innovation process (Fransman 2010) (see Table 1).

Table 1. Six symbiotic relationships in the ICT ecosystem

Relationship
1	Relationship between network element providers and network operators.
2	Relationship between network operators and content and applications providers.
3	Relationship between content and applications providers and final consumers.
4	Relationship between network element providers and final consumers.
5	Relationship between network element providers and content and applications providers.
6	Relationship between network operators and final consumers.

Source: Fransman (2010)

Fransman (2010) also addresses the ecosystem and the implications of public policies to foster innovation. For him, the following systemic factors have a marked influence on innovation in the ICT sector:

- sectoral regulation (mainly for network operators) and antitrust legislation. For Fransman (2010), the intensity of competition and cooperation among firms is closely associated with innovation;
- availability of financial resources (ease and flexibility to obtain financial resources), costs of loans and equity. Essentially, capital and inherent risk play a key role in the approach;
- other R&D agents, such as universities, research centres, standardisation forums, and intellectual property protection entities; and
- government support.

Diga and May (2016) propose a conceptual framework for the ICT ecosystem, based on contributions from previous works, studying the interactions between these groups of key actors, the presence of infomediaries facilitating the interpretation and use of information, and considering factors of gender, generation, ethnolinguistics, historical context, accessibility, access and use shaping the diverse relationships of people within the ICT system.

These authors also consider the institutional arrangements, national policies and governance structures guiding the use of ICT by the various society members. As a result of this structure, human well-being must also be analysed as a result of the ICT system and human interaction, with the possibility of being improved, stagnant or degraded, leading to changes in physical, social, economic and political aspects. The objective of this conceptual framework is to simultaneously capture technical, economic, and institutional relationships in an emerging domain that is not yet sufficiently mature for adequate modelling with more specialised methods, hence its application in the context of the open government data ecosystem.

Open government data value chain

The generation and use of open government data (OGD) comprises a chain with a set of different combinations of activities such as production, collection, publication, enrichment and organisation of data, access, use, distribution, update, visualisation, services, among others activities, in which each activity can add value to the data for potential users through the production of information and services (Davies 2014; Pellegrini 2012; Robinson et al. 2009; Ubaldi 2013).

One step of the OGD value chain is the identification and classification of each actor, to understand their role in the chain. According to Ubaldi (2013), in order to analyse the OGD value chain, it is essential to understand the role of the main actors and the speed with which the context evolves, causing significant changes in the roles and limits of their responsibilities, thus affecting the actors' involvement in the chain activities.

The sequence in which these activities can occur in the value chain will be defined by the needs of the potential users and the aggregation of value provided by each actor. For example, the agency that produces and publishes the data can be responsible for several of these activities such as production, organisation, publication, and data updating. Previous research has classified the actors according to their roles in the execution of these activities (Davies 2014; Deloitte Analytics 2012; Magalhães et al. 2014; Scott 2014; Ubaldi 2013) which are summarised as follows:

- **Suppliers** collect and produce data as a result of the institution's mission.
- **Publishers** organise and prepare data in open standards and formats for publication.
- **Enrichers** increase the value of data by checking, cleaning and elaborating data, adding links or new data.
- **Aggregators** collect, integrate and analyse open government data and data from other sources that allow them to meet specific information needs.
- **Enablers** add value, by treatment, selection, analysis, and integration of open government data, to their products and services; to meet the specific needs of their users.
- **Distributors** collect, organise and redistribute information, such as repositories and portals.
- **Facilitators** provide platforms or tools that facilitate access to or use of government open data.
- **Developers** build and distribute applications for specific purposes.
- **End-users** (organisations, individuals or government) use government open data for their own economic, social or environmental activities.

Ubaldi (2013) proposes the organisation of the value chain of OGD in four phases, grouping a delimited set of activities, based on a study of the value chain for Public Sector Information in the European Community (Dekkers et al. 2006):

- **Data generation:** at this stage the data are generated. In general, this activity is carried out by public sector entities or by entities financed with public resources;
- **Data collection, aggregation, and processing:** at this stage data are collected and organised in order to allow access, sharing and reuse, being elaborated, connected or combined to add value to potential users;
- **Data distribution and delivery:** at this stage, data are distributed to potential users for access; and
- **Final data use:** at this stage, data are reused by different users. The larger and more varied these groups of users, the greater the potential for sustainability of the ecosystem and the indication that the value adding to the data has been successful.

Creating value in OGD along the value-chain implies an adequate understanding of the needs of potential users, in terms of content and format, and also an adequate understanding of how data are created, protected, shared and used (Ubaldi 2013). The final use of OGD can occur in several ways. Davies (2010) identified at least five processes of OGD use, which may occur in a distinct or combined manner, according to the purpose of the use.

- **Facts understanding (*Data to fact*):** When a newly opened dataset is used to look for specific facts. These facts may be related to civic engagement, business planning, or understanding bureaucratic processes. In this use, the data would be submitted to research activities, careful analysis, and extraction of information;
- **Information production (*Data to information*):** when one or more datasets are used to construct a representation and interpretation of a given subject, leading to the production of infographics or reports. In this use, the data are submitted to activities such as manipulation, statistical analysis, visualisation, contextualisation and elaboration of reports;
- **Interface availability (*Data to interface*):** when one or more datasets are interactively accessed or exploited through interfaces that allow the analysis and formulation of interpretations without the need for advanced technical knowledge for the data manipulation. In this use, the data are submitted to activities such as cleaning, combining, sub-division and display in preconfigured interfaces;
- **Data production (*Data to data*):** when a dataset is treated for the production of secondary data, in order to allow access in a new format, to complement it with other data or other use. This includes activities such as base format conversion, filtering, complementation and combination, and providing access to the resulting database; and

- **Production of services (*Data to service*):** when data are used for the production of services. In this use, the data can be submitted to a great number of activities in order to incorporate them into already existing or the creation of new products and services.

Ubaldi (2013) describes how OGD affect value creation for different ecosystem actors, such as government, citizens, civil society, and the private sector, where a market for services is developed from the open data. According to this author, value creation from OGD should address issues such as increasing the amount of public data available in open digital format, exploring and understanding the barriers to the process of data entry at various levels of government, data ownership, the role of the various actors involved, and the assessment of the return on investment to identify the facilitators of value creation. The researcher also emphasises that understanding and fostering the use of OGD is critical to the value creation process as well as to the creation of a sustainable and collaborative ecosystem in data creation.

Jetzek et al. (2014) propose a structure composed of four prototypes of mechanisms of value creation, showing how OGD reuse can stimulate value creation in a sustainable manner, that consider not only economic aspects with a focus on increasing revenue and profit but also new aspects of economic impact such as resource sharing and the collaborative production of goods and services.

In this model mechanisms are organised in two dimensions: the first, analysing the value generated by the use of data to improve existing products and services or the provision of new products and services, and the exploration of new opportunities; the second, identifying whether reuse is the result of a market exploration in the traditional economic format, with the economic value generated by reducing costs and increasing revenues, or in a shared economic format, based on a network economy of organisations interacting with each other.

Another model proposed to understand how data-driven innovation emerges from OGD and how this innovation creates social and economic value for society is presented by Abella et al. (2017) in a context of intelligent cities. This model addresses the different stages of the process, from the acquisition of data to its processing for the creation of services and the analysis of their impact.

The first stage focuses on the characterisation of the data according to their degree of reuse. This characterisation is an input to the second stage and allows correlating the available data with the innovative products and services developed in this stage. The second stage focuses on how these data sources, through reuse, become innovative services and products that are used even by the organisation that generates the data. The third stage assesses the economic and social impact generated by these products and services. The economic impact may be to increase incomes or decrease costs for existing activities. The social indicators would be related to the involvement, experience, security, and confidence of the users (Abella et al. 2017).

Governance in relationships between organisations

Governance is the process by which organisations (social actors) influence their environment in order to create order and continuity in the social system (Gomes & Merchán 2017). According to Rasche and Gilbert (2012: 101)

> the overall aim of governance is to create the conditions for social order by emphasising the need for the regulation of social affairs. Regulation [...] can take many different forms, including self-regulation of private actors, co-regulation of public and private actors [...].

Examples of regulation mechanisms are rules and/or standards developed by public and private actors and state legislation (e.g. laws of command and control drawn up by states) (Gomes & Merchán 2017).

There is also the special case of transnational governance focusing on the forms of regulation that emerge and are not operated exclusively by the state (Cashore 2002). The concept of transnational governance arises from contemporary capitalism and the transnational relations that emerge from it, such as global production and value chains, the flow of information and capital between countries, and the financialisation of everyday relations (i.e. greater dependence on the financial sector for the realisation of present and future consumption, such as the private pension fund and other forms of investments, credit card and mortgages) (Djelic & Sahlin-Andersson 2006, in Gomes et al. 2017; Morgan 2016, in Gomes et al. 2017; Morgan & Kristensen 2012, in Gomes et al. 2017).

According to Gomes et al. (2017), the regulation process takes place mainly through interested organisations and system participants, whose interactions and negotiations are constantly creating regulatory arenas. The basic activity is not only to elaborate rules and certifications, but also to monitor and audit their implementation. Even without the elaboration of new rules or certifications, negotiation can already have a regulatory effect. Transnational governance can take two forms: (a) soft law or soft regulation: self-regulated certifications and standards (e.g. involving only pro-market actors) or co-regulation (e.g. when there are public and private actors); (b) hard law: the law and its forms of command and control, defining what is permitted or not in a given jurisdiction (Djelic & Sahlin-Andersson 2006; Rasche & Gilbert 2012).

Regardless of the form taken by the transnational governance, three characteristics or dimensions can be highlighted: (a) it involves multiple actors, both the state and its agencies, as well as civil society organisations and companies; (b) it is immersed in multilevel relationships, that is, the local, national and global dimensions intertwine and it is often difficult to separate analytically what happens in each of them; and (c) negotiation, since forms of governance are not the result of the imposition of a single actor (such as the state), but are the result of negotiation and consensus building on the rules and norms that will influence the behaviour of the actors. Regarding this last item, it can be said that governance involves steering, since regulation is constructed and negotiated through the exercise of influence (Djelic & Sahlin-Andersson 2006; Rasche & Gilbert 2012).

In business environments, the concept of meta-organisations (MO) defines the type of organisational structure where the members themselves are organisations, which, based on shared decisions and interests, such as social interest or standardisation, may constitute a model of partial organisation, in which some of the criteria that define a complete organisation (definition of members, hierarchy, rules, monitoring, and accountability) are not met in their entirety (Ahrne & Brunsson 2010).

Gulati et al. (2012) studied variations in MO architectures according to the model adopted for membership participation and for internal decision-making, leading to a definition of dimensions and taxonomies that includes the networks of organisations or individuals not linked by authority or labour relations, but by common goals. König et al. (2012) have shown that MO, similar to other organisations, may respond with inaction when faced with disruptive innovations, due to MO characteristics such as decision-making format, political organisation, and organisational complexity, and observed that these characteristics may increase or decrease inertia to the response.

Berkowitz and Dumez (2015) studied MOs in the oil and gas sector and found that the companies organised their own environment through the creation of different types of MOs, with limited dimensions and performance, when faced with characteristic situations. One such situation is the need to manage relationships with multiple actors, such as civil society and governments (see Figure 1).

Figure 1. Three situation decisions that originate MO

Source: Berkowitz & Dumez (2015)

With many redundant MOs subsisting and coexisting, Berkowitz and Dumez (2015) argue that the business environment is affected by increasing organisational complexity due to this proliferation of MO. Finally, the study also emphasises the importance of membership of MOs, their implications for legitimacy issues and the emergence of strategy.

Methodology

This study started with a literature review for the definition of the theoretical reference model for the OGD ecosystem analysis, selecting the concepts of value chain for OGDs, ICT ecosystems, meta-organisations and governance of ICT ecosystems.

Jabareen (2009) suggests that the construction of a framework from existing multidisciplinary literature occurs through a process of theorising, which is based on the methodology of grounded theory rather than a simple description of the data and the phenomenon studied. Thus, the author defines a framework as a network, or a 'plan', between concepts that together offer a comprehensive understanding of the phenomenon or of phenomena within a context.

These concepts, within the framework, support each other, playing ontological, epistemological and methodological roles. The ontological role of concepts within the framework is linked to the knowledge of the 'path of things', 'the nature of reality', 'real' existence, and 'real' action. The epistemological role is linked to 'how things really are' and 'how things really work' in a known reality. Finally, the methodological role is linked to the process of constructing the framework and how it can tell us 'something' about the 'real' world (Guba & Lincoln 1994).

Preliminary results

Considering the objectives of the study, the concepts presented in the theoretical framework, and the methodology, reached the main elements of the proposed framework for OGD ecosystems analysis. These elements were organised and resulted in the following table:

Table 2. Synthesis of the concepts for the construction of the framework

Concepts	Concept description	Categorisation
Open government data and OGD projects	• The culture of government openness adopted and disseminated; • OGD public policies made official; • the diffusion of standards and capacities; and • standards and technologies adopted.	epistemological
ICT ecosystem applied to OGD	• Identify and characterise the role of intermediaries; • the organisation and management of the network of actors involved; • agencies and government entities that fill gaps in the value chain; • business models adopted by intermediaries and users; and • value creation made possible by the chain.	ontological

Concepts	Concept description	Categorisation
OGD value-chain	• Identify and characterise the role of information intermediaries; • the organisation and management of the network of actors involved; • agencies and government entities that fill gaps in the value chain; • business models adopted by intermediaries and users; and • value creation made possible by the chain.	methodological
Meta-organisation theory	Define the type of organisational structure where the members themselves are organisations; understand the possible variations in meta-organisation architecture; and how the structure responds to internal or external stimuli.	ontological
Transnational governance	Ways to exercise regulation in meta-organisations; how new members can be incorporated by the meta-organisation and how they can leave the meta-organisation; and how transactions resulting from relationships between actors and ecosystem levels are managed.	ontological

Source: Elaborated by the authors

An OGD ecosystem, in a conceptual analogy with the biological ecosystems observed in nature, is a system that includes all living organisms (e.g. OGD producers, intermediaries and final users) in an area, as well as its physical environments (e.g. public policies, legal environment, platforms, data value chain, system governance) functioning together as a unit. It can be characterised by one or more equilibrium states, where a relatively stable set of conditions exists to maintain actors and value creation and exchange at desirable levels. The ecosystem should have certain functional characteristics that specifically regulate change or maintain the stability of a desired equilibrium state.

An OGD ecosystem models the economic dynamics of the complex relationships that are formed between actors or entities whose functional goal is to enable use and reuse of OGD. In this context, the actors would include the material resources (data, systems, platforms, etc.) and the human capital (data producers and publishers, intermediaries, users, etc.) that make up the institutional entities participating in the ecosystem (e.g. the government, non-profit organisations, business firms, university research institutes, and economic development and business assistance organisations, funding agencies, policymakers, etc.).

The OGD ecosystem comprises three distinct economies, the data collection economy, which is driven by governmental function, the production economy, which is driven by actors and entities engaged in making data published, and the commercial economy, which is driven by the marketplace and by economic value generation of OGD through services and products offered to final users. Figure 2 illustrates the actors, the data and exchange value flows and the rules and legal arrangements involved in the OGD ecosystem.

Figure 2. OGD ecosystem

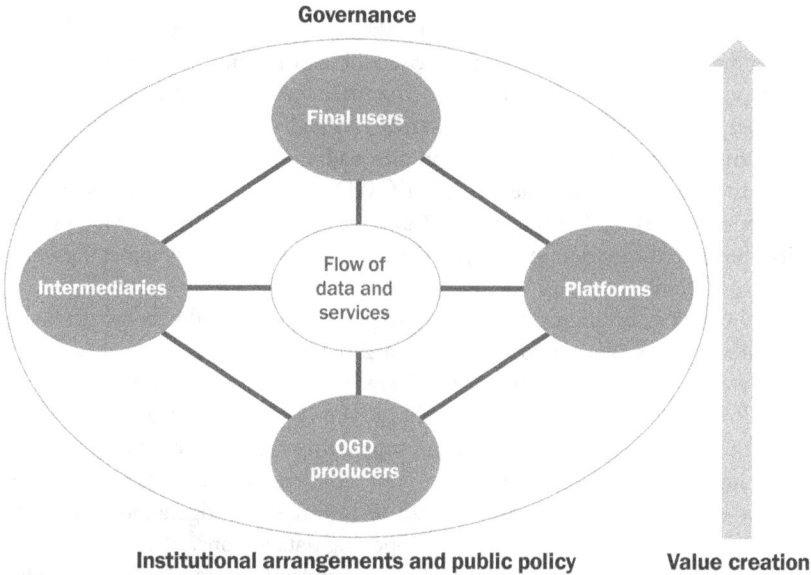

Source: Elaborated by the authors, adapted from Enhanced ICT ecosystem framework (Diga & May 2016)

With the concepts selected for the construction of the OGD ecosystem analysis framework (see Table 2), it will be possible to contribute to an analysis and understanding of the governance necessary for the construction and sustainability of OGD ecosystems so that society can obtain greater benefit in the use of this information as well as how government resources (allocated in these projects) are used more efficiently, with a view to maximising results.

It is expected that the proposed framework will work as a reference for future studies that seek to understand, for example, what the conditions for the creation and maintenance of an economically and socially sustainable OGD ecosystem are or what factors affect the creation and maintenance of an ecosystem formed by producers, intermediaries, and users, where value is added to the OGD throughout the activities of the chain generating information flow, products, services and financial resources.

The identification of studies of open data projects, without any of them having sufficient scope of actors and organisations, in order to guide the factors that give legitimacy to the OGD ecosystem, was considered as one of the limitations of this work. The validation of the framework, to be carried out in future studies, will be an important stage in the support and credibility of the proposed framework and will also subsidise the *rethinking* of the framework due to the inherent dynamics of the ecosystem.

Discussion and conclusion

The proposed model is being used in the analysis of the Brazilian federal government OGD ecosystem, comprised of legislation and public policies, strategy, governance and operational structure, federal government and other agencies contributing data, infomediaries and end-users, citizen participation mechanisms. The OGD ecosystem comprises a large number of internet portals, the principal ones being the Brazilian OGD portal, the main access to the system and the Brazilian transparency portal and the OGD monitoring portal, with the special function of assuring government transparency and compliance with regulations.

The OGD ecosystem began being implemented in 2004, being formally established in its present form by law in 2012 and considered now to be an already consolidated structure with a large intra-government and citizen user base. Preliminary analysis results indicate the importance of a central actor monitoring the compliance of laws and supporting the agencies in creating and executing their open data plans. The main challenges in the implementation process were related to government staff technical knowledge about open data standards and resources to funding the opening data actions.

The analysis demonstrates the importance of understanding OGD as a complex ecosystem, with regulations and governance recognising the multiple public and private actors and their continually evolving relationships. This becomes especially important in assuring the system's sustainability in administrative transitions.

REFERENCES

Abella A, Ortiz-de-Urbina-Criado M & De-Pablos-Heredero C (2017) A model for the analysis of data-driven innovation and value generation in smart cities' ecosystems. *Cities* 64: 47–53 https://doi.org/10.1016/j.cities.2017.01.011

Ahrne G & Brunsson N (2010) Organization outside organizations: The significance of partial organization. *Organization* 18(1): 83–104. https://doi.org/10.1177/1350508410376256

Berkowitz H & Dumez H (2015) How firms (partially) organize their environment: Meta-organizations in the oil and gas industry. *i3 Working Papers Series* No. 15- CRG-02. https://hal.archives-ouvertes.fr/hal-01483012

Cañares MP (2014) Opening the local. Paper presented at ICEGOV '14: Proceedings of the 8th International Conference on Theory and Practice of Electronic Governance, Guimaraes, Portugal, October 2014. https://doi.org/10.1145/2691195.2691214

Cashore B (2002) Legitimacy and the privatization of environmental governance: How non-state market-driven (NSMD) governance systems gain rule-making authority. *Governance* 15(4): 503–529. https://doi.org/10.1111/1468-0491.00199

Chattapadhyay S (2014) Access and use of government data by research and advocacy organisations in India: A survey of (potential) open data ecosystem. Paper presented at ICEGOV '14: Proceedings of the 8th International Conferences on Theory and Practice of Electronic Governance, Guimaraes, Portugal, October 2014. https://dl.acm.org/doi/abs/10.1145/2691195.2691262

Davies T (2010) Open data, democracy and public sector reform: A look at open government data use at data.gov.uk. Practical Participation. http://practicalparticipation.co.uk/odi/report/wp-content/uploads/2010/08/How-is-open-government-data-being-used-in-practice.pdf

Davies T (2014) *Open data in developing countries – Emerging insights from Phase I, (2)*. Berlin: World Wide Web Foundation. http://www.opendataresearch.org/sites/default/files/publications/Phase 1 – Synthesis – Full Report-print.pdf

Dekkers M, Polman F, Te Velde R & De Vries M (2006) *Measuring European public sector information resources: Final report of study on exploitation of public sector information-benchmarking of EU framework conditions*. Brussels: HELM & Zenc. https://ec.europa.eu/digital-single-market/en/news/mepsir-measuring-european-public-sector-information-resources-final-report-study-exploitation-0

Deloitte Analytics (2012) *Open growth stimulating demand for open data in the UK – A briefing note from Deloitte Analytics*. London. https://www2.deloitte.com/content/dam/Deloitte/uk/Documents/deloitte-analytics/open-growth.pdf

Diga K & May J (2016) The ICT ecosystem: The application, usefulness, and future of an evolving concept. *Information Technology for Development* 22 (sup.1): 1–6. https://doi.org/10.1080/02681102.2016.1168218

Djelic M-L & Sahlin-Andersson K (2006) Transnational governance in the making – Regulatory fields and their dynamics. In: M-L Djelic & K Sahlin-Andersson (eds) *Transnational Governance: Institutional dynamics of regulation*. Cambridge: Cambridge University Press. pp. 1–47. http://regulation.upf.edu/ecpr-05-papers/mdjelic.pdf

Fransman M (2007) Innovation in the new ICT ecosystem. *Communications & Strategies* 68: 89–110

Fransman M (2010) *The New ICT Ecosystem – Implications for Policy and Regulation*. New York: Cambridge University Press. www.cambridge.org/9780521171205

Gomes MVP & Merchán CR (2017) Governança transnacional: Definições, abordagens e agenda de pesquisa. *Revista de Administração Contemporânea* 21(1): 84–106. https://doi.org/10.1590/1982-7849rac2017150332

Guba EG & Lincoln YS (1994) Competing paradigms in qualitative research. In: NK Denzin & YS Lincoln (eds) *Handbook of Qualitative Research*. Michigan: SAGE Publications. pp. 105–117. https://doi.org/http://www.uncg.edu/hdf/facultystaff/Tudge/Guba%20&%20Lincoln%201994.pdf

Gulati R, Puranam P & Tushman M (2012) Meta-organization design: Rethinking design in interorganizational and community contexts. *Strategic Management Journal* 33(6): 571–586. https://doi.org/10.1002/smj.1975

Gurstein MB (2011) Open data: Empowering the empowered or effective data use for everyone? *First Monday* 16(2). https://doi.org/10.5210/fm.v16i2.3316

Intarakumnerd P & Chaoroenporn P (2013) The roles of intermediaries in sectoral innovation system in developing countries: Public organizations versus private organizations. *Asian Journal of Technology Innovation* 21(1): 108–119. https://doi.org/10.1080/19761597.2013.810949

Jabareen Y (2009) Building a conceptual framework: Philosophy, definitions and procedure. *International Journal of Qualitative Methods* 8(4): 49–62. https://doi.org/10.1177/160940690900800406

Janssen K (2011) The influence of the PSI directive on open government data: An overview of recent developments. *Government Information Quarterly* 28(4): 446–456. https://doi.org/10.1016/j.giq.2011.01.004

Janssen M, Charalabidis Y & Zuiderwijk A (2012) Benefits, adoption barriers and myths of open data and open government. *Information Systems Management* 29(4): 258–268. https://doi.org/10.1080/10580530.2012.716740

Janssen M & Zuiderwijk A (2014) Infomediary business models for connecting open

187

data providers and users. *Social Science Computer Review* 32(5): 694–711. https://doi.
org/10.1177/0894439314525902

Jetzek T, Avital M & Bjørn-Andersen N (2014) Generating sustainable value from open
data in a sharing society. *IFIP Advances in Information and Communication Technology*
429: 62–82. https://doi.org/10.1007/978-3-662-43459-8_5

Johnson JA (2014) From open data to information justice. *Ethics and Information
Technology* 16(4): 263–274. https://doi.org/10.1007/s10676-014-9351-8

König A, Schulte M, & Enders A (2012) Inertia in response to non-paradigmatic
change: The case of meta-organizations. *Research Policy* 41(8): 1325–1343. https://doi.
org/10.1016/j.respol.2012.03.006

Magalhães G, Roseira C & Manley L (2014) Business models for open government data.
Paper presented at ICEGOV '14: Proceedings of the 8th International Conference on
Theory and Practice of Electronic Governance, Guimaraes, Portugal, October 2014.
https://doi.org/10.1145/2691195.2691273

Magalhães G, Roseira C & Strover S (2013) Open government data intermediaries. Paper
presented at ICEGOV '13: Proceedings of the 7th International Conference on Theory
and Practice of Electronic Governance, Seoul, Korea, October 2013. https://doi.
org/10.1145/2591888.2591947

Mittilä TS (2006) Intermediary organisation in a regional development network. In:
The Proceedings of the 22nd IMP Conference, Porvoo -Borgå, Finland. pp. 1–10. http://
citeseerx.ist.psu.edu/viewdoc/download?doi=10.1.1.494.126&rep=rep1&type=pdf

Pellegrini T (2012) Semantic metadata in the news production process – Achievements
and challenges. MindTrek '12: Proceeding of the 16th International Academic
MindTrek Conference, Tampere, Finland, October 2012. https://doi.org/10.1145/
2393132.2393158

Rasche A & Gilbert DU (2012) Institutionalizing global governance: The role of the
United Nations global compact. *Business Ethics: A European Review* 21(1): 100–114.
https://doi.org/10.1111/j.1467-8608.2011.01642.x

Roberts A (2014) Making transparency policies work: The critical roles of trusted
intermediaries. Legal Studies Research Paper Series 14–39. Boston, MA. http://
www.freedominfo.org/2014/10/making-transparency-policies-work/%0Ahttp://www.
freedominfo.org/2014/10/making-transparency-policies-work/

Robinson D, Yu H, Zeller WP & Felten EW (2009) Government data and the invisible
hand. *Yale Journal of Law & Technology* 11: 160–175. http://ssrn.com/abstract=1138083

Scott A (2014) *Open data for economic growth*. Washington, DC: World Bank Group.
http://documents.worldbank.org/curated/en/131621468154792082/Open-data-for-
economic-growth

Sein MK & Furuholt B (2012) Intermediaries: Bridges across the digital divide.
Information Technology for Development 18(4): 332–344. https://doi.org/10.1080/02681
102.2012.667754

Ubaldi B (2013) Open government data: Towards empirical analysis of open government
data initiatives. *OECD Working Papers on Public Governance* No. 22. Paris: OECD
Publishing. www.oecd.org/daf/inv/investment-policy/working/papers.htm

Van Schalkwyk F, Cañares M, Chattapadhyay S & Andrason A (2016) Open data
intermediaries in developing countries. *The Journal of Community Informatics*
12(2). https://figshare.com/articles/Open_Data_Intermediaries_in_Developing_
Countries/1449222

Van Schalkwyk F, Willmers M & Czerniewicz L (2014) Case study: Open data in the
governance of South African higher education. http://www.opendataresearch.org/
project/2013/uct

World Bank (2016) *World Development Report 2016: Digital Dividends*. Washington,
DC: World Bank Group. https://doi.org/10.1596/978-1-4648-0671-1

10.

From theory to practice: Open government data, accountability and service delivery

Michael Christopher Jelenic

Over the past decade, the production of data has grown exponentially around the world, and now more than ever, complex data are available at the touch of a button – from the grand, such as DNA sequencing and macroeconomic modelling, to the more mundane, such as weather maps, trash collection schedules, and public transport timetables. Expanding at a rapid pace, experts estimate that there will be a further 4 300 per cent increase in annual data generation, reaching a level of 35 zettabytes (equivalent to 35 trillion gigabytes) by 2020.[1] As the sheer amount of available data in the world increases – driven by advances in technology and computing – terms such as 'big data' and 'open data' have become common parlance, reflecting the importance of data use in our everyday lives.

In particular, open data – that is, 'digital data that is made available with the technical and legal characteristics necessary for it to be freely used, reused, and redistributed by anyone, anytime, anywhere' (Open Data Charter 2015: 1) – has attracted much attention as a means to innovate, add value, and improve outcomes in a variety of sectors. Based on its attributes of 'availability and access, reuse and redistribution, and universal participation' (Open Knowledge Foundation 2012), it has been argued that open data has the potential to deliver a number of benefit streams. In particular, open data's potential to foster innovation, efficiency, jobs, profits, and economic growth has been touted by the private sector, which has widely emphasised the potential financial and economic returns of open data. For instance, McKinsey estimates that open data can help

---

1 See "Big Data Universe beginning to explode," at http://www.csc.com/insights/flxwd/78931-big_data_universe_beginning_to_explode

unlock between USD 3 trillion to USD 5 trillion in economic value annually across seven sectors, including education, transportation, consumer products, electricity, oil and gas, healthcare, and consumer finance (Manyika et al. 2013: 9–12). Likewise, previous work by the World Bank has suggested that open data can yield economic benefits across five archetypical types of businesses, including suppliers, aggregators, developers, enrichers, and enablers of open data (Stott 2014). Other economic analyses of the topic include: a European Union study that estimated aggregate direct and indirect economic benefits of €200 billion (Vickery 2011); a UK study that 'conservatively' estimated the direct economic benefits of public sector information at around £1.8 billion a year (Shakespeare 2013); and a study produced in Spain that found that the 'infomediary' sector employs around 4,000 people in the country and generates €330 million to €550 million annually (Tinholt 2013).

Compared with the attention given to these financial and economic benefits – however mind-bogglingly large they are purported to be – considerably less effort has been made in providing a sufficiently critical evaluation of the public-sector benefits, including the potential of *open government data* (OGD). As a subset of open data, OGD is that which is 'held by national, regional, local, and city governments, international governmental bodies, and other types of institutions in the wider public sector' (Open Data Charter 2015: 3). As defined by the OECD, open government data can be seen as comprised of two elements: (i) government data, which are 'any data and information produced or commissioned by public bodies'; and (ii) open data, which 'are data that can be freely used, re-used and distributed by anyone, only subject to (at the most) the requirement that users attribute the data and that they make their work available to be shared' (Ubaldi 2013: 6). Common types of OGD include, but are not limited to, data on maps, land ownership, census, government budget, government spending, company registration, legislation, public transport, international trade, health, education, crime statistics, environment statistics, election results, and contracts.

While there is considerable conjecture about the precise nature and categorisation of public sector benefits that can be derived from these types of data, OGD has been tied to a number of positive public externalities. For its part, the World Bank has cited a number of benefit streams, including: (i) fostering economic growth and job creation; (ii) improving the efficiency, effectiveness, and coverage of public services; (iii) increasing government transparency, accountability, and citizen participation; and (iv) facilitating better information sharing within government itself (World Bank 2015b). Likewise, the OECD describes similar benefits, which include: (i) improving government accountability, transparency, responsiveness and democratic control; (ii) promoting citizens' self-empowerment, social participation and engagement; (iii) building the next generation of empowered civil servants; and (iv) creating value for the wider economy (Ubaldi 2013).

Recognising this potential, dozens of governments have created OGD platforms to support the proactive release of high-value datasets that can be used to improve government transparency, accountability, and performance. Notable examples include the United States and the United Kingdom, who pioneered the release of government datasets in an effort to make government more transparent through platforms such as data.gov and data.gov.uk. Since then, numerous countries have created similar OGD portals to provide a plethora of government information (Farhan, D'Agostino & Worthington 2012). At a global level, international institutions and non-governmental organisations (NGOs) have begun providing technical assistance and capacity building for open data initiatives around the world. In particular, the drive toward greater government transparency has been supported by the Open Government Partnership (OGP), which was formed in 2011 to function as 'an international platform for domestic reformers committed to making their governments more open, accountable, and responsive to citizens.'[2] Likewise, in order to deepen commitments to government transparency, an Open Data Charter was signed in 2013 by the G-8 countries, setting out the principle of 'open by default,' which has subsequently prompted calls for a 'data revolution' to take centre stage in the post-2015 development agenda, including the dialogue on the Sustainable Development Goals. Other notable initiatives include the Open Data for Development Partnership (OD4D), comprising 65 bi-laterals, multilaterals, foundations, and NGOs; the Open Data Working Group of the OGP; the Open Contracting Data Partnership; and the Global Partnership for Sustainable Development Data; as well as tools such as the Open Data Barometer, which provides information on data availability in more than 90 countries.

Given the proliferation of international initiatives to support Open Government, and OGD in particular, it is increasingly evident that OGD is a key concern for political leaders in developing and developed countries alike. As such, the motivation for initiating and promulgating such initiatives can be viewed as a product of public policy priorities at the national level, which may differ significantly across countries. For developed countries, OGD is often guided by domestic legislation related to freedom of information (FoI) and access to information (A2I) as well as fiscal pressures, which make it imperative to demonstrate improved administrative efficiency and value for money in service delivery. For developing countries, these same motivations exist, but are amplified by a number of related issues, including, inter alia, an interest to satisfy donors and attract international aid; to meet international commitments such as the Sustainable Development Goals (SDGs); and to improve perceptions of transparency and accountability, which can help to improve the business climate and attract foreign investment.

---

2   See https://www.opengovpartnership.org/

Problem statement

Based on its potential promise, a murky theory of change seems to have emerged among open data advocates, practitioners, and 'evangelists,' which suggests that OGD has the potential to make governments more transparent in their operations and performance, thus providing an evidentiary basis for citizens to hold politicians to account – in a principal-agent manner – for spending public resources and providing public services. However, despite the appeal of this logic, there is very limited cross-country empirical data to support the purported relationships between these issues. In its place, it would appear that advocates have increasingly turned to a 'faith based' view on the efficacy of OGD, which has a number of limitations. One strand of literature on the efficacy of OGD comes in the form of 'success stories' drawn from different settings around the world. Such qualitative case study analysis – if it can be labelled as such – tends to express the effects of open data in 'aspirational and speculative' terms, often providing little evidence that open data 'outputs' such as policies, platforms, dashboards, and the like are actually impacting public sector 'outcomes' such as improved accountability, service provision, or service outcomes (Verhulst & Young 2017). Moreover, the situation specificity of these cases – often referring to a particular government, in a particular country, in a particular sector, at a particular moment in time – provides limited external validity for other contexts.

Likewise, from a quantitative perspective, there is a significant body of impact evaluation literature utilising randomised controlled trials, as well as quasi-experimental research designs, which has found statistical associations between government transparency initiatives and improved government accountability and service delivery. However, while this strand of literature has evaluated government transparency and accountability initiatives (TAIs) more broadly – including fiscal transparency programmes, community scorecards, and access to information legislation, among others – few, if any, impact evaluations have been performed on the treatment effects of OGD initiatives in particular. Moreover, almost all of these impact evaluations are done at the country or sub-national level, focusing on very specific populations and specific moments in time. As with the qualitative case study literature described above, the existing quantitative evidence therefore has a number of spatial and temporal limitations, which reduce its applicability to other country contexts.

Research question and methods

Given the limitations of the available current research, whether qualitative case studies or quantitative impact evaluations, it appears that a significant gap exists in the literature between theory and practice. In an attempt to fill this lacuna, the present research asks: based on the evidence currently available, to what extent and for what reasons is the use of OGD associated with higher levels of accountability and improved service delivery in developing country contexts?

To begin to answer this question, this research will first attempt to determine whether a statistical association actually exists to support the purported relationships between the variables. In doing so, the research will rely on a unique dataset that operationalises OGD, government accountability, and service delivery – including both provision and outcomes – while operationalising other intervening variables that capture critical conditions for improved accountability and service delivery. Given the paucity of comparative studies on this topic, this research will utilise cross-country data for over 25 countries in sub-Saharan Africa over a time horizon for which data currently exist.

Significance and contributions

The present research aims to investigate the relationship between OGD, accountability, and service delivery in order to provide one of the first statistical assessments of the available evidence in a cross-country comparative context. Importantly, it is not the intent of the present research to provide a *definitive* or *causal* assessment of the relationship among OGD, accountability, and service delivery. Rather, given the limitations of the data available, the empirical contribution of this research is to flesh out potential correlations between variables in order to provide an *initial* assessment of OGD's efficacy. In doing so, this research attempts to identify key indicators – which operationalise a conceptual framework gleaned from the literature – to test what associations may exist between variables. Moreover, the empirical contribution of this research seeks to tease out and test the effects of a number of intervening variables, as suggested by the literature, in order to have a fuller sense of the political economy dynamics at play.

In addition to this empirical contribution, however small, this research also seeks to make a contribution to the theoretical literature by providing a conceptual framework, which captures current discussions and debates on OGD, transparency, accountability, and service delivery. In addition, this research seeks to contribute to the collective understanding of the necessary and sufficient conditions that support OGD, government accountability, and improved service delivery more broadly. In particular, this research hopes to demonstrate the enabling role that access to information and political participation can play in terms of leveraging sources of OGD to hold governments answerable and accountable to citizens. Likewise, in political environments where these factors are legally and institutionally constrained, this research seeks to test alternative hypotheses with respect to public service delivery, including the managerial role played by bureaucrats themselves as well as transparency and social accountability movements more broadly.

193

A final practical contribution of this work is to provide a better understanding of how OGD is released and utilised in developing country contexts with respect to accountability and service delivery. As such, the countries included in this assessment were chosen given that OGD has the potential to more significantly 'move the needle' with respect to accountability and service delivery in these environments than in high-income country contexts, where sophisticated accountability institutions and service delivery structures are already in place. Finally, and perhaps most significantly, the focus of this research on developing countries is especially salient as these often represent the weakest governance environments, which are in most need of improvements to public accountability and service delivery institutions.

Literature review

The central research question proposed above sits at the intersection of a wide range of theoretical and empirical literature relevant to modern public-sector management and international development. In particular, this research touches on previous work related to the delineation of 'big data,' of which 'open data' and 'open government data' are extensions. Taken further, OGD as a concept can be viewed as a more specific element of a larger issue area related to government transparency and accountability initiatives (TAIs), which may take a myriad of manifestations including, among others, open budgeting, asset disclosure, and declassification initiatives, as well as freedom of information (FOI) and access to information (A2I) legislation. For its part, the literature on transparency is often tightly linked with literature on government accountability, including key theoretical concepts such as democratic accountability, social accountability, and social capital as well as those related to citizen voice, participation, and engagement. Underpinning these conceptions of accountability is a large body of literature on public management and administration, including that related to new public management, public choice theory, and welfare economics, which provides a theoretical foundation for analysing the public provision of social services.

Given the diversity of themes surveyed, the literature review that follows is presented in two parts. The first element is a theoretical section, which provides an overview of the key concepts, including open government data, transparency, accountability, and service delivery. Given that there is significant conjecture among authors on the exact meaning of these terms, this first section seeks to provide a plurality of perspectives, as well as to specify the definitions that will be used throughout this analysis, including the conceptual framework put forth in the following section. The second part of the literature review presents the available qualitative and quantitative empirical literature, which highlights the current understanding of the relationships between OGD, transparency, accountability, and service delivery.

194

Theoretical perspectives, definitions and differentiations

The primary topic of this research – open government data (OGD) – touches on a very active and ongoing contemporary debate on the boundaries of a number of concepts which have emerged over the past decade. While not entirely scholarly in its nature, numerous governments, international institutions, and private sector entities have proffered differing definitional propositions with respect to terms such as 'big data,' 'open data,' and 'open government data,' albeit with significant overlap and interconnection. Accordingly, as a first step, it is necessary to disentangle the various concepts to best specify a clear, working definition of OGD, which is the main topic of this research.

Distinctions in terminology have been provided by various sources, with differing levels of conceptual specificity. For instance, McKinsey and Co. refer to 'big data' as datasets that are 'voluminous, diverse, and timely,' with modifier 'big' referring to the size and complexity of these sets. 'Open,' in McKinsey's definition, describes how liquid the data are – that is, how 'transferable' these data are across platforms and users (Manyika et al. 2013). Relatedly, the World Bank's research on the topic posits that the term 'big data' describes very large and complex datasets that must be processed using advanced analytic techniques, whereas the key defining characteristic of 'open data' is that it is made available as a public good (World Bank 2016). These distinctions are related to those of other authors, such as those of the Open Data Barometer, who describe knowledge 'as "open" if anyone is free to access, use, modify, and share it – subject, at most, to measures that preserve provenance and openness.' Drawing on these notions, the Open Data Institute likewise highlights centrality of shareability and access in its definition of open data as '[d]ata that is made available by governments, businesses, and individuals for anyone to access, use, and share' (Open Data Institute 2015a). However, a more nuanced definition comes from the Open Data Handbook, which delineates a number of functions that data must fulfil to be deemed as 'open'. As such, open data must possess three key attributes:

- **Availability and Access:** Data must be available as a whole and at no more than a reasonable reproduction cost, preferably by downloading over the internet.
- **Reuse and Redistribution:** Data must be provided under terms that permit reuse and redistribution including the intermixing with other datasets.
- **Universal Participation:** Everyone must be able to use, reuse, and redistribute; there should be no discrimination against fields of endeavour or against persons or groups. (Open Knowledge Foundation 2012)

Other authors have taken more parsimonious perspectives on the relationship among big data, open data, and OGD, which speaks to the nuance of the concepts and further elucidates the subject of this research. Gurin (2014: 13) notes that big data are usually passively generated, and often kept private, including 'data that retailers keep on customers' buying habits, that cell phone companies keep on their mobile users, or that hospitals collect about their patients.' Alternately, Gurin characterises open data as 'public' and 'purposeful' – and as such, 'information that has a particular goal in mind, such as fuelling new businesses, improving public health, or identifying wasteful spending.' Further illustrating the overlap and interaction of these terms, the World Bank likewise notes the following: 'Open data are those that are freely and easily accessible, machine-readable, and explicitly unrestricted in use. Open data aren't necessarily big, and big data aren't necessarily open' (World Bank 2016).

Drawing on these conceptions, the Open Data Charter, which was first signed in July 2013, further delineates the difference between open data, and what constitutes *open government data*, which is the central concept of the present research. In particular, the Charter 'recognises that the term "government data" includes, but is not limited to, data held by national, regional, local, and city governments, international governmental bodies, and other types of institutions in the wider public sector.'[3] Given the plurality of perspectives on these closely interrelated topics, the present research will rely on – for the sake of simplicity and clarity – a two-part definition of OGD as put forth by the OECD (Ubaldi 2013). Such a conception marries two constituent elements: (i) 'government data,' which includes 'any data and information produced or commissioned by public bodies;' and (ii) 'open data,' which, similar to other definitions, 'can be freely used, re-used and distributed by anyone, only subject to (at the most) the requirement that users attribute the data and that they make their work available to be shared as well' (Ubaldi 2013: 6).

The current research question also relates to a growing literature on transparency and conceptions of 'openness' in government. Similar to what constitutes OGD, transparency has no single definition nor is there a single indicator to measure it. Instead, transparency takes on a variety of different meanings and measures across organisations and within the literature. Khemani et al. (2016: 5) provide one of the simplest and most straightforward definitions of transparency as 'citizen access to publicly available information about the actions of those in government and the consequences of those actions.' Similarly, Trapnell (2014) operationalises transparency as greater availability of information between government departments as well as greater clarity about government processes, rules, and definitions. Somewhat more colourfully, Transparency International, one of the world's leading authorities on government transparency, describes it as such:

Transparency is about shedding light on rules, plans, processes and actions. It is knowing why, how, what, and how much. Transparency ensures that

---

3   See https://opendatacharter.net/history/

public officials, civil servants, managers, board members and businesspeople act visibly and understandably, and report on their activities. And it means that the general public can hold them to account. It is the surest way of guarding against corruption and helps increase trust in the people and institutions on which our futures depend.[4]

For the purposes of the current research, Khemani's definition of transparency is perhaps most useful in that it delineates two distinct elements – one in providing information about performance, and another about the outcomes of that performance. As such, Khemani (2016: 220) notes that 'information provided through transparency must be specific about both policy actions and the resulting outcomes, so that citizens can use this information to select and sanction leaders.' In the context of OGD and service delivery, this would amount to having information not only on what was budgeted for public services, such as health and education, but also how and to what extent those budgets were executed, what was the quality and quantity of the services provided, and what were the eventual outcomes in terms of human development.

The idea of having information about both actions and outcomes is similarly captured by Fox (2007), who argues that transparency can be either 'clear' or 'opaque'. Fox differentiates between these two conceptions of transparency, with 'opaque' transparency referring to 'the dissemination of information that does not reveal how institutions actually behave in practice, whether in terms of how they make decisions, or the results of their actions (2007: 667).' In the context of OGD, access to information (A2I) and freedom of information (FoI) legislation may enable the public to access large datasets and a government may even proactively disclose this information; however, not all of this information may provide insights into government performance or policy outcomes, thus making it 'opaque'. A case in point would be a 'data-dump' where an enormous quantity of data may be disclosed, but the quality, accessibility, and usefulness of the data are limited. Fox (2007: 667) contrasts such 'fuzzy' forms of transparency with 'clear' transparency, which is 'reliable information about institutional performance, specifying officials' responsibilities as well as where public funds go.' Linking this to Khemani's (2016) example above, 'clear' transparency would likewise include information about budgeted resources, actual spending, provision, and related outcomes.

A critical element of this research is to determine how and to what extent OGD – as a transparency initiative in and of itself – makes government more accountable. This question touches upon the historical aspects of Open Government and OGD, including its linkage to transparency and accountability. As such, Yu and Robinson (2012) argue that some level of 'ambiguity' has emerged with respect to OGD, noting that the concept no longer refers only to policies that support improved government accountability. Rather, 'the

---

4   http://www.transparency.org/ what-is-corruption/#what-is-transparency

term "open government data" might refer to data that make the government as a whole more open (i.e. more publicly accountable), or instead might refer to politically neutral public sector disclosures that are easy to reuse, even if they have nothing to do with public accountability' (Yu & Robinson 2012: 1). As a result of this 'new ambiguity,' Yu and Robinson argue that the release of open data, or the creation of special websites and OGD portals, may be enough to deem a government 'open' – but may not actually make it more transparent or accountable. This distinction between the terms – as well as the distinctions related to 'clear' and 'opaque' transparency noted above – suggest that researchers cannot assume a merely linear relationship between these concepts. Rather, as noted by Kosack and Fung (2014), these relationships are often intermediated by a number of contextual factors that determine the extent to which information can effectively inform individual choices, assist with collaborative problem solving, increase pressure on service providers, enable top-down reforms, as well as support countervailing power structures.

The central research question engages with the debate on what constitutes accountability, including public sector accountability more specifically. However, just as transparency and OGD have a wide variety of interpretations, the concept of accountability is equally challenging to define. A very broad and useful definition is provided by Tisné (2010), who notes that:

> Accountability refers to the process of holding actors responsible for their actions [...] it is the concept that individuals, agencies, and organisations (public, private, and civil society) are held responsible for executing their powers according to a certain standard (whether set mutually or not).

Building on the concept of accountability as a process, there is considerable debate in the literature regarding how these mechanisms function over time and with respect to different actors. Using a principal-agent framework, the World Bank has conceptualised accountability as a set of relationships among actors (e.g. individuals, agencies, and organisations) – both public and private – responsible for their actions and executing their powers according to a certain standard (World Bank 2003). Such a rationale assumes that principals – namely, citizens – delegate their sovereignty to elected representatives to undertake certain public governance responsibilities as well as to allocate the resources – in the form of taxes and transfers – to pay for these services. As agents, these actors – namely, politicians, policymakers, and providers – are responsible for performing the delegated tasks and to inform the principals about their performance. Based on this information, the principals can then enforce sanctions and rewards on the agents, both through electoral and non-electoral channels.

Challenging the functioning of this principal-agent relationship, the World Bank's 2017 *World Development Report* (2016) identifies a number of 'power asymmetries,' which may emerge. These include: (i) 'exclusion,' whereby certain individuals and groups may not enjoy the same access to certain services; (ii) 'capture,' whereby elite individuals or groups influence politicians to adopt policies that serve a narrow special interest, often at odds with the public good; and (iii) 'clientelism,' whereby political support is exchanged for preferential access to certain goods and services. Similar aberrations are noted by Devarajan and Khemani (2016), who describe how such forms of 'unhealthy political engagement' can similarly 'invert' the principal-agent relationships between citizens and government officials.

Other authors note that government transparency and government accountability are often conflated at the conceptual level, and there is a general predisposition to assume that transparency automatically leads to improved accountability. To take on this misconception, numerous authors have argued that while governments can become more 'transparent,' they do not automatically become more 'accountable,' unless there are appropriate mechanisms in place for that data to be acted upon by either the government or citizens more broadly. To be sure, Fox (2007) argues that accountability can be either 'soft' or 'hard,' which reflects two critical dimensions: 'answerability,' that is, the right/capacity to demand answers, as well as 'enforceability,' with the capacity to sanction, compensate, or remedy. The underlying relationships alluded to in this description are also touched upon by Peixoto (2013), who further argues that 'publicity' and 'political agency' conditions are necessary to translate transparency initiatives, such as OGD, into greater public accountability.

Other useful conceptions of accountability related to the present research come from the larger political science scholarship on democratic accountability. Conceptions of the citizens ceding their sovereignty to elected officials and holding them accountable through electoral and administrative channels can be traced back as far as the work of Alexis de Tocqueville (1956), who saw these democratic accountability mechanisms as the greatest checks against despotism and tyranny. However, while foundational, this early work only began to lay the foundations of more complex notions of how accountability functions in democratic societies. More recently, Dahl and Lindblom (1976) emphasise the convergence of a number of fundamental processes of control, including market systems, democratic institutions, bargaining among leaders, as well as effective control of non-leaders through hierarchical relationships. Likewise, Borowiak further defines the attributes of democratic accountability with reference to the principle that 'the governed should have the opportunities to sanction and demand answers from the powers that govern them' (2011: 9). A common theme uniting all of this related literature is an implicit recognition of the principal-agent relationship within accountability processes.

Taking an international perspective, other authors have written more broadly on the role of democratic accountability in the organisations – such as international development institutions, multilateral transparency initiatives, and other global charters and regimes – which support global agendas such as OGD and transparency. Dahl (1999) examines the extent to which these organisations can likewise ensure the appropriate levels of democratic accountability demanded from their members and clients. In particular, the 'democratic deficits' that can emerge when delegated responsibilities make it difficult for citizens at the domestic level to exercise effective control over larger 'bureaucratic bargaining systems.' Likewise, Grant and Keohane (2005) provide a typology of accountability mechanisms, including hierarchical, supervisory, fiscal, legal, market, peer, and reputational – which are more relaxed and nuanced than the strict democratic accountability relationships that function at the domestic level. Similarly, Keohane (2003) differentiates between accountability mechanisms, which may influence the way that international agendas (e.g. OGD) are managed in terms of authorisation, support, and impact. While these arguments are very salient in terms of the debates on democratic accountability at the international level, they only tangentially touch upon the type of 'domestic' democratic accountability referred to in this present research, which essentially is located at the national and sub-national levels where services are delivered.

Moving from accountability relationships to the service delivery mechanisms themselves, one of the most significant analyses of a principal-agent conception of service delivery is provided by the World Bank's 2004 *World Development Report* (World Bank 2003), which developed a 'Service Delivery Framework' comprised of four constituent elements (citizens/clients, politicians/policymakers, organisational providers, frontline professionals). Recognising that public intervention is necessary in the provision of social services, a 'long-route' to accountability and service delivery would involve citizens – acting as principals – using their 'voice' to demand improvements in services from policymakers, who serve as agents. For their part, policymakers rely on 'compacts' with service provider organisations, such as ministries of health and education or regional hospitals or school districts. In this 'long-route', provider organisations then 'manage' the frontline service providers – the doctors, nurses, teachers, and engineers – to more effectively deliver services. A final step in this accountability chain is the role of 'client power' – a condition in which end-users express their preferences and levels of satisfaction to frontline providers.

While such a model of service delivery is rooted in conceptions of democratic accountability, it can likewise be contextualised with respect to the growth of the welfare state since the end of the Second World War and the emergence of new public management theory in the early 1990s. Both of these phenomena have created an environment where the role of the government has changed to become, respectively, both a provider and manager of public goods such as social services. From a demand side, the rise of the welfare state largely explains the growth of social service provision. To be sure, Myles and Quagagno (2009) argue that since the 1960s, expenditures on social transfers have grown considerably as a result of social insurance schemes as well as national social service programmes in health and education. From a supply side, new changes in government management techniques have put a premium on delivering services in an accountable, efficient, and evidence-based manner.

These debates on the production and allocation of public services, particularly in the context of a growing welfare state, are captured in the vast literature on public choice theory, including the work of, inter alia, Black (1958), Arrow (1951), Downs (1957), Buchanan and Tullock (1962) and Olson (1965). As such, the underlying rationale is that the production of services will be dictated by a confluence of interests and actors within the political sphere, whereby self-interest ultimately determines the allocation of resources. As such, certain goods such as health and education, may be undersupplied by market forces, necessitating the government to step in where the private market cannot – or will not – provide them. The idea of differentiating types of goods, including public goods that may be undersupplied due to market failures, can be linked to the early work of Samuelson (1954: 387), who defined 'collective consumption goods' as those 'which all enjoy in common in the sense that each individual's consumption of such a good leads to no subtractions from any other individual's consumption of that good.' Conceptions of 'non-rivalry' and 'non-excludability' in describing public goods and services are further emphasised by the work of Ostrom (2005), who treats the production of goods and services as a matter of public choice and bargaining based on diverse individual incentives. Marrying the work of Samuelson and Ostrom, OGD in this context can be seen as a collective consumption good, which is both non-excludable and non-rivalrous.

It is precisely at this junction where the present research on OGD links to public administration theory, including new public management (NPM), which recognises government's special role in service provision, financing, and regulation (World Bank 2003). As first defined by Hood (1991), NPM has been postulated to provide for better provision of public services, particularly in contexts of heightened public accountability, declining fiscal space, and focus on performance results. In particular, Gruening (2001) highlights the diverse antecedents of NPM, as a mixture of public choice theory, managerialism, policy analysis, principal-agent theory, property rights theory, transaction cost economics, and methodological individualism, which allows for a more flexible, outcome-driven approach to public sector management. As such, key characteristics of NPM include, among others: accountability for performance, performance auditing, privatisation, decentralisation, competition, performance measurement, freedom to manage, and increased use of information technology. In the context of OGD, NPM theory is especially relevant because it provides the basis for performance measures – in terms of outcomes and not just programme inputs – that can help determine the efficacy of service delivery as well as allow for evidence-based decision-making and subsequent action based on managerial discretion.

Importantly, it should be noted that while not incompatible, NPM departs from the classical public administration theory, which emphasises the role of bureaucratic inputs managed by a top-down power structure. As such this classical public administrative theory – as first put forth by Max Weber's seminal 'Bureaucracy' (originally published in 1922) – makes a case for a top-down control mechanism in the form of a 'monocratic hierarchy' whereby policy is formulated at a high level and then executed by a series of descending offices. Within this context, Weber described the role of the civil servant as 'vested in his ability to execute conscientiously the order of the superior authorities' (2015: 95). This classical model is also a hallmark of 'Taylorism' – based on Taylor's *Principles of Scientific Management* – which emphasised the importance of control and planning mechanisms to best ensure efficiency and accountability. Although NPM does move away from these classical public administration perspectives, it provides an important antecedent for the conceptual model described in the next section, particularly, the ability of government to manage public servants to deliver services.

Empirical evidence and links between concepts

Current empirical evidence of the effects of OGD remains somewhat sparse; however, efforts have been made to study the relationship between OGD and accountability, albeit with mixed results. As such, Davies (2014) challenges an implicitly linear theory of change and argues that policymakers need a more nuanced understanding of how open data leads to outputs, outcomes, and impact. Using qualitative case studies, Davies finds that while there is evidence of open data outputs such as applications, dashboards, and new analysis, information on outcomes is limited. Nevertheless, his case study review highlights the importance of legal frameworks, domain knowledge, and technical skills as important background resources for reaching outputs such as applications, reports, analysis, and new derived datasets (World Bank 2003). Similarly, other case study research on open data's impacts in Kenya's urban slums and rural settlements suggests that data are being used by these communities and that they are demanding even greater quantities of data; however, the study finds no clear association between open data and improved accountability or service delivery.

A similar stream of research investigates the role of transparency initiatives more broadly, particularly the role played by Freedom of Information (FoI) and Access to Information (A2I) legislation – which have the ostensible goal of improving government accountability. As such, empirical studies have suggested a connection between improved accountability and government responsiveness, although the linkage between FoI/A2I and accountability is less clear. For instance, it has been postulated that FoI initiatives, particularly those that are community-based, have been linked to improved citizen rights related to housing and water in South Africa. Similarly, Hazell, Worthy, and Glover (2010) provide a systematic evaluation of FoI laws in Britain, finding that such legislation can help improve government decision-making and public awareness. Likewise, Peisakhin and Pinto (2010) analyse FOI legislation in India and find that it has a significant impact in motivating government officials to process applications for certain entitlement programmes, including reducing the time needed to collect benefits (Hazell, Worthy & Glover 2010). Pandey et al. (2008a) find that FOI has reduced absentee rates of public-school teachers in rural India, albeit with limited effect on eventual learning outcomes in the schools. Finally, Awortwi and Nuvunga (2019) investigate the role that information disclosure regimes in the extractives sector – such as the Extractives Industry Transparency Initiative (EITI) – can play in improving accountability. Surveying 17 different empirical factors, the authors find that unless there is a risk of a ruling party losing power, information disclosures are unlikely to make governments more accountable on their own.

Relatedly, there is a relatively large literature that supports the proposition that the users of OGD – for instance the press, academia, and civil society – can use this information to hold governments to account. For instance, Mungiu-Pippidi (2014) investigates – among other variables – press freedom and corruption. Using a large-n cross-country analysis, as well as many of the same corruption indexes used by this present paper, the author finds a robust correlation between the variables. Similarly, Ferraz and Finan (2008) find that in Brazil, the proactive disclosure of audit outcomes significantly impacted electoral performance of incumbents, particularly when corrupt practices were uncovered. Interestingly, these effects seem to be amplified in places where public radio was active in transmitting this information, which suggests the importance of the media in perpetuating the accountability enhancing effects of OGD. Similar results were obtained in Mexico by Larreguy, Marshall, and Snyder (2014), who studied the performance of infrastructure grants that targeted the poor. The authors found that when provided information on the use and outcomes of these funds, voters tend to punish poorly performing politicians, but only in areas covered by local media – again, suggesting the key role that the media can play in disseminating information to hold government to account.

In addition to investigating the role of OGD, the present research also seeks to build on previous efforts (Joshi 2013) investigating the efficacy of transparency and accountability initiatives, the findings of which are mixed in terms of service delivery. To be sure, Gaventa and McGee (2013: s16) have evaluated the efficacy of a number of transparency and accountability mechanisms,[5] and note that where there is 'positive evidence in one setting, this is often not corroborated – and sometimes even contradicted – by findings in another setting where different, or even similar, methods have been used.' Likewise, Carothers and Brechenmacher (2014) note that evidence on the impact of accountability,

---

5   These include, among others, citizen report cards, organisational score cards, and asset disclosures of public officials, as well as more 'participatory' mechanisms such as community monitoring, social audits, public expenditure tracking surveys, and participatory budgeting processes.

transparency, participation, and inclusion is 'often long-term, indirect, and difficult to isolate from other factors, and the evidence base to date is still too thin to arrive at firm conclusions.'

Nevertheless, there is a growing body of large-n impact evaluation literature which provides some empirical evidence of transparency and accountability initiatives impacting service delivery, particularly in the education sector. For instance, Reinikka and Svensson (2004, 2011) have estimated that the dissemination of budgetary information has led to less leakage of public funds in Uganda under certain conditions, and Barr et al. (2012) have found that participatory monitoring has been linked to improved education outcomes. Similarly, Pandey et al. (2008b) provide empirical evidence from India which suggests that dissemination of budget information has improved teacher effort including reduced absenteeism. In a seminal study in Kenya, Duflo, Dupas and Kremer (2012) have similarly demonstrated a positive association between community-based hiring of teachers with improved attendance and test scores.

Just as in the education sector, there is a growing body of empirical evidence that similarly links transparency and accountability initiatives with improved provision and outcomes in the health sector. Björkman and Svensson (2009) and Nyqvist, De Walque & Svensson (2014) have provided evidence that access to service delivery information and related outcomes can have an effect on how services are delivered. Using a randomised controlled trial in the Uganda health sector, the studies found that better service delivery and outcomes were associated with the treatment group that received a report card on staff and health centre performance. Similarly, Misra and Ramansakar (2007) have demonstrated that community scorecards have improved satisfaction in health service delivery in India, which suggests the importance of community members having objective and quantitative information regarding staff behaviour in monitoring accountability. Additional work on these issues has likewise been completed by Ravindra (2004), who found that citizen report cards actually improved local service delivery in certain settings in India, and Goldfrank (2006) who demonstrates that participatory budgeting initiatives have been shown to improve public service delivery in Brazil.

With respect to electoral accountability mechanisms, the present research seeks to confirm earlier evidence, which supports the proposition that elections can have a significant effect on citizens rewarding and sanctioning politicians for their performance in delivering services. To be sure, there is a significant body of literature that suggests that performance in service delivery is an important metric that voters use to choose politicians, and to vote out of office those whom they believe are performing poorly. Using the Italian municipal level to test this hypothesis, Kendall et al. (2015) find that information on 'valence' issues, such as past service delivery performance, increased support for candidates much more than information about their particular policy positions. Similar work by Carruthers and Wanamaker (2015) finds that the enfranchisement of women in elections during the early part of the 20th century in the United

States correlated with an increase in public school expenditure, suggesting that formal electoral mechanisms can have an effect on service provision.

Such findings have likewise been found in the health sector, where enfranchisement of the poor has been shown to have a significant effect on health provision and outcomes. Fujiwara (2015) finds that the rise of electronic voting in Brazil resulted in the de facto enfranchisement of poor and less educated voters, which had the secondary effect of increasing healthcare provision, including services for prenatal and newborn health. Similarly, in terms of health outcomes, Kudamatsu (2012) utilises surveys conducted in 28 African countries to compare infant survival rates before and after the emergence of democracy in these countries and finds that infant mortality declined after democratisation, which suggests the effect that formal democratic accountability channels may have on improving health outcomes, particularly in a developing country context.

Gaps in the literature

While the theoretical literature reviewed above provides an understanding of how the concepts of OGD, transparency, accountability and service delivery are defined – and how they relate to larger theoretical concepts and debates – the empirical literature that supports a relationship between these ideas is slightly less complete. As noted above, existing case studies are often aspirational or speculative in nature, and do not provide sufficiently rigorous evidence that OGD inputs translate into accountability or service delivery outcomes. Moreover, the situation specificity of these cases provides limited external validity for other contexts. A final drawback to this literature is a paucity of analysis of the proximate conditions in place – that is, defining what is necessary and sufficient from a political economy perspective – to ensure that OGD has its intended impacts.

Likewise, with respect to the empirical literature, numerous studies have been completed – some using sophisticated large-n experimental and quasi-experimental designs – however, few have been able to link open data, and OGD in particular, to changes in levels of government accountability or the provision and outcomes of health and education services. Rather, the literature available focuses mainly on the effects of other types of transparency and accountability initiatives – such as community score cards, participatory budgeting, and social audits – and not OGD itself. As is the case with such granular quantitative analysis, many studies focus on particular countries and regions at particular moments in time, which again, reduces their comparability to other contexts.

Conceptual framework

In an attempt to answer the question posed above, this research will evaluate four different propositions – or premises – that can be extrapolated from the literature, including previous theoretical and empirical contributions on transparency, accountability, and service delivery. As described below, this research will also test the political economy 'conditions' that are necessary and/or sufficient for the purported relationships to hold – including the role played by the larger information availability environment (i.e. 'publicity' condition) and formal citizen voice (i.e. 'political agency' condition) as well as government capacity (i.e. 'public sector management' condition) and informal social accountability channels (i.e. 'client power' condition).

Premise 1: Higher levels of OGD make government more accountable

Under this premise, OGD provides an evidentiary basis through which citizens can be better informed on how government is performing. In particular, for this information to be effective – for it to be a type of 'clear' transparency as noted by Fox (2007) – two elements would need to be present. In line with Khemani's (2016) findings, this would first include information on sectoral performance, both in terms of provision of services (e.g. number of teachers, number of vaccinations, etc.) as well as service delivery outcomes (e.g. completion rates, infection rates, etc.). Second, in order for the efficiency of such service provision and outcomes to be determined, transparent information is needed both on how much was budgeted for these services as well as how much was actually spent on these services (e.g. execution rate). Only by having both types of data can 'demand-side' consumers – that is, citizens, civil society organisations, researchers, journalists, and other 'infomediaries' – make informed judgements as to whether the government is adequately executing its mandate and meeting its responsibilities.

However, it should be noted that transparent access to sectoral and budgetary data is just one element; in order to hold government officials accountable, citizens need mechanisms to exercise their 'voice', either formally or informally. In line with the earlier work of Fox (2007), such voice channels could provide a degree of 'answerability' and 'enforceability', which is crucial to ensuring a principal-agent element in the citizen-state interface. Formal citizen voice channels would include electoral mechanisms in the form of free and fair elections, which are held at regular intervals in line with legal provisions, and with no *de jure* or *de facto* impediments to participation. Other non-electoral, or social accountability mechanisms, could include, inter alia: ICT enabled platforms[6] for collecting citizen feedback, satisfaction surveys, complaint and redress mechanisms, petitions, protests, and other participatory institutions for citizens.

Given the need for both information to be publicly available as well as for mechanisms to be in place for citizens to express their voice and hold service providers to account, the research will investigate a number of enabling

---

6   Examples of such ICT platforms include Maji Voice in Kenya; I Change My City in India; Pressure Pan in Brazil; and U-Report in Uganda.

'conditions,' which have been postulated by the earlier work of Peixoto (2013), as noted above. These include the following:

- **Publicity condition:** A vital element necessary for supporting citizen voice is the 'publicity condition,'[7] which provides for an enabling environment whereby 'disclosed information actually reaches and resonates with its intended audiences'. Conceptualised as such, the publicity condition would presuppose the existence of a necessary degree of political freedoms and civil rights so that the disclosed data can reach citizens – if not directly, then through third-party intermediators such as CSOs, social movements, academics, and the media. At minimum, this condition would imply, inter alia: constitutional protections for freedom of expression, speech, press (and internet); relevant freedom of information or access to information legislation; media freedom, impartiality, and protections against censorship; as well as the ability for civil society organisations, academia, unions, and other civic organisations to operate free from harassment, discrimination, or persecution.

- **Political agency condition:** A second critical element is the presence of a 'political agency condition,'[8] which provides the legal and institutional 'mechanisms through which citizens can sanction or reward public officials.' Conceptualised as such, the most obvious mechanism for citizens to express their voice would be through free and fair elections; however, this condition would imply the existence of other non-electoral mechanisms for public participation, including civil participation in political parties, CSOs, NGOs, as well as other participatory institutions. Evidence that this condition is in place would include, inter alia: evidence of political participation through free and fair elections; necessary oversight and electoral monitoring mechanisms; as well as political pluralism, including freedom of political parties, CSOs, and NGOs from discrimination, or persecution.

In order for this premise to be true, there would be a number of observable implications. First, we would expect to see higher levels of OGD produced in terms of sector performance data (e.g. in health and education) as well as higher levels of data on budget allocation and execution. Second, we would expect government accountability indicators to likewise increase over time, with a significant correlation to improvements in OGD. Finally, in line with the literature,

---

7   This condition relates to other concepts in the literature such as 'answerability', 'soft' accountability, and 'yelp'.

8   This condition relates to other concepts in the literature such as 'enforceability', 'hard' accountability, and 'teeth'.

we would expect publicity and political agency conditions to be positively and significantly correlated with increases in accountability. This premise would be falsified if either there was an inverse relationship between OGD and government accountability, or, indicators measuring publicity and political agency were not positively correlated with OGD and government accountability.

Premise 2: More accountable governments enable improved service delivery

According to this second premise, policymakers and elected representatives would be held accountable to respond to citizens 'voice' – whether formally or informally expressed – or risk the consequences of inaction. As such, these policymakers must be able to discipline government agencies that execute the public policy directives, priorities, and initiatives that they are advocating. Moreover, each service delivery agency must be able to ensure that civil servants – including the frontline providers responsible for service delivery – are adequately rewarded/sanctioned for their performance in carrying out their delegated tasks.

While publicity and political agency are needed to translate OGD into an accountability tool, policymakers must likewise have the capacity to actually enact the policies they are seeking to implement. However, in many developing country contexts, it is often the case that a significant 'implementation gap' emerges between *de jure* legislation, policies, or regulations and *de facto* policy implementation. While impossible to generalise across all developing countries, such implementation gaps tend to emerge in two particular instances. The first is when politicians are not able to effectively delegate to line ministries – as their agents – to carry out policy implementation. In many developing country contexts, this is the result of competing mandates, poor coordination, unclear reporting relationships, poor prioritisation, and limited institutional capacity. The second instance where implementation gaps may emerge, is at the level of the frontline service providers – that is, the doctors, nurses, teachers, engineers, sanitation workers, who physically interact with citizens in providing services. Policy implementation at this level is also prone to failure if the necessary resource allocation, human resource capacities, and performance incentives are not in place. For instance, in a context where high absenteeism is prevalent and disciplinary measures are not enforced, it is doubtful that service delivery provision and outcome targets will be reached.

Returning to the literature, both of these phenomena have been well documented empirically and theoretically. With respect to the relationship between the politicians and the line ministries they oversee, previous research highlights the importance of 'compacts' with service provider organisations, such as ministries of health and education or regional hospitals or school districts. As such, policymakers are capacitated as principals, who provide resources and delegate power to these provider organisations and have the capacity to reward or sanction their agents if the need arises. With respect to the relationship between line ministries and frontline service providers, it is necessary that provider organisations effectively 'manage' the frontline service providers – the doctors, nurses, teachers, engineers, and so forth – using the appropriate sanctions, rewards, and incentives.

Therefore, according to this premise – where higher levels of government accountability lead to higher levels of service delivery – we would expect to see a well-functioning system of institutional 'compacts' as well as strong 'management' capacity. In order to capture these two somewhat fuzzy concepts, the research will investigate the effects of an additional 'condition' as follows:

- **Public management condition:** In this step of the accountability chain, policymakers function as the principals to demand that their agents – namely, service delivery organisations, such as ministries of health and education – are taking the necessary actions to improve services. In addition, service provider organisations must be able to effectively manage their frontline service providers, who in turn serve as their agents. As noted above, in order to operationalise this chain of principal-agent relationships, governments rely on 'compacts' and 'management', which would imply a number of processes, procedures, and institutional capacities being in place, including, inter alia: revenue mobilisation and taxation capacity; capacity for adequate public financial management, including fiscal policy formulation, budget management, and multiyear planning; statistical capacity; as well as general public administration capacity, including HR and performance management mechanisms.

In order for this premise to be true, there would be a number of observable implications. First, we would expect to see higher levels of government accountability corresponding with improved provision of services as well as improved outcomes in key social sectors, such as health and education. Second, we would expect that improved levels of service delivery are associated with increasing levels of public management capacity, similarly demonstrating a positive and significant association. Accordingly, this premise would be falsified if either there was an inverse relationship between levels of government accountability and service delivery, or, indicators measuring the strength of public management capacity were not positively correlated with improved service delivery.

Premise 3: Higher levels of OGD make governments more accountable, which subsequently enables improved service delivery

Often referred to as the 'long-route' to service delivery, this third premise unites the earlier premises into an accountability chain that links citizens, to government, to providers in a series of principal-agent relationships. As such,

OGD would provide an evidentiary basis for citizens to hold their politicians to account for their mandates, and these politicians would then exercise 'compacts' over line ministries to develop and implement policies for service delivery, and these ministries would then be able to discipline frontline service providers such as doctors, nurses, and teachers, through good 'management.' Crucial to this hypothesis is the role that accountability plays as an *intermediating variable*, namely that OGD improves accountability, and *because of this improved accountability*, service delivery is improved.

In order for this third premise to be true, we would expect OGD, accountability, and service provision and outcomes to have a significant, positive relationship. At the same time, we would expect the joint effects of OGD and accountability (i.e. the interaction effect) to have a higher and more significant coefficient than either that of OGD or accountability on their own. Such a statistical association would demonstrate that accountability intermediates the relationship between OGD and service delivery, thus supporting this premise. Accordingly, this premise would be falsified if either there was an inverse relationship between levels of OGD, government accountability, and service delivery – and no intermediating effects is found for accountability at the sectoral level.

Premise 4: Higher levels of OGD directly enable improved service delivery, excluding the effects of accountability

Under this fourth premise, formal accountability channels are bypassed, and a direct link is assumed between OGD and service delivery. As such, this direct route can be 'bottom-up' (upward) or 'top-down' (downward) in nature.

With respect to the former, clients – namely citizens, with the help of CSOs and infomediaries – would have access to OGD that provide vital information on public performance. In such a case, access to OGD can provide the evidentiary basis through which citizens can make determinations about the quality of services in local hospitals and schools. Based on this information, citizens then communicate their preferences directly with frontline service providers, including the teachers, doctors, and other direct service delivery actors. For instance, in Madagascar and other developing country contexts, the use of parent–teacher committees provide a valuable monitoring mechanism for parents as well as a mechanism for parents to communicate with education service providers about the quality and content of the services they provide. As demonstrated in the literature, citizens may rely on other social mechanisms as well, such as community scorecards, participatory budgeting exercises, and social audits, in order to directly influence the quantity and quality of services provided.

'Bottom-up' (upward) channel: In order for such a channel to be evident, citizens would not express their voice through traditional accountability mechanisms such as elections or political processes, but do so in a more direct manner, interfacing directly with local service providers. In line with the conceptual model advocated by the World Bank's 2004 World Development Report, the research will investigate the effects of a final condition, as follows:

- **Client power condition:** In contexts where there are weak electoral mechanisms or weak government management capacity, client power can be especially helpful in collectively organising local communities. In this scenario, characterised by limited, formal political accountability mechanisms, it would be necessary for there to be an ecosystem in place that allows clients to supervise the provision of services as well as to communicate their preferences, satisfaction, and complaints directly with service providers themselves. Such a condition would imply mechanisms similar to the civil transparency and social accountability mechanisms mentioned above, including participatory budgeting, citizen score cards, social audits, as well as parent-teacher councils and local health centre supervisory councils. For these to be successful in expressing citizen voice to frontline providers, a number of crucial elements would need to be in place. First, this would entail a strong presence and capacity of civil society organisations, which can organise around key issue areas, monitor performance, collectively represent constituent voices, and communicate communal level concerns directly with frontline providers, for instance at schools and hospitals. This condition would also presume a level of IT, internet connectivity, mobile phone penetration, and data literacy, if not on the part of individual citizens, then by CSOs, journalists, researchers, or other 'infomediaries,' who play a role in digesting and disseminating information to the general population. Finally, on a larger level, the functioning of strong client power relationships would rely on a baseline of political and civil rights, which would permit freedom of association, media, and speech as well as a legal framework that protects civic groups and the media from harassment and censorship.

In order for this premise/channel to be true, there would be a number of observable implications. First, we would expect to see higher levels of OGD corresponding to improved service delivery. In this relationship, citizens, CSOs, and other 'infomediaries' would be the direct consumers of this data and exercise their voice outside of formal accountability mechanisms. As such, we would expect to find a well-developed civil society ecosystem, the capacity for the sharing of data at the grassroots level, and social accountability mechanisms in place to allow citizens to engage providers directly. Accordingly, this premise would be falsified if either there was an inverse relationship between levels of OGD and service delivery, or the control variable operationalising the 'Client Power' condition was not significant.

'Top-down' (downward) channel: Under such a scenario, changes to government services can be made unilaterally based on the strength of public management capacity (e.g. 'public management condition' as described above) in order to obtain whatever exigencies and incentives managers deem most important and feasible. While this hypothesis has not been widely advocated in the literature, numerous studies suggest that OGD is able to facilitate better information sharing within government. To be sure, in places where OGD portals have been introduced, a large proportion of the views or downloads are from civil servants themselves. Examples of such behaviour may be supported by non-democratic, 'managerialist,' or 'developmentalist' regimes, which have shown a proclivity for implementing service delivery reforms based on solid evidence, yet without electoral incentives to respond to citizen accountability pressures.

In this premise, citizen voice – or direct 'client power' – is removed, and service delivery improvements are predicated solely upon internal government power relationships and government capacity to command and control public servants. In place of 'client power' and 'citizen voice' are strong government 'compacts' (described above) as well as a high degree of 'management' in the form of sanctions and reward mechanisms to discipline or reward frontline service providers. In this step of the accountability chain, policymakers function as the principals to demand that their agents – namely, service delivery organisations, such as ministries of health and education – actually take action to improve services.

In order for this premise/channel to be true – we would expect to see higher levels of OGD lead to improved service delivery, irrespective of 'client power' and formal political accountability channels. In their place, we would expect to see a significant and positive association with the 'Public Management' condition described above. This premise would be falsified if either there was an inverse relationship between levels of OGD and service delivery, or the control variable operationalising the 'Public Management' condition was not significant.

Figure 1. Theoretical framework of key premises

Research design and methods

With respect to the statistical model, this research utilises ordinary least squares (OLS) multivariate regression equations to capture the associations between the variables. As such, in premise 1, the key independent variable is OGD and the regression specification aims to test its relationship on accountability, controlling for the Publicity condition, Political Agency condition, and log GDP. In premise 2, the key independent variable is accountability and the regression specification aims to test its relationship on service delivery – including both provision and outcomes – controlling for the Public Management condition and log GDP. Building on these first two regressions, in premise 3, the key independent variable is OGD and the regression specification aims to test its relationship on service delivery – including both provision and outcomes – controlling for the intermediating effects of accountability as well as log GDP. Finally, in premise 4, the key independent variable is OGD and the regression specification aims to test its relationship on service delivery – including both provision and outcomes – controlling for the Public Management condition, the Client Power condition, and log GDP.

Sample selection

Based on the most recently available data, the sample will include 25 countries in sub-Saharan Africa[9] over a four-year period, 2013–2016. As such, the unit of analysis is country by year (e.g. Ghana 2014, Tanzania 2016, etc.) for a total of 86 data points. These countries were chosen because they are at reasonably similar levels of human development and per capita income, which allows for good comparability between cases. Likewise, the country sample presents relatively good variation with respect to the key variables over the period, allowing the research to better isolate key associations. Finally, the country selection will allow comparison at the sub-regional level, as key differences may emerge between the southern/eastern cluster versus the western cluster of African countries.

Primary variable/indicator construction

To perform these regressions, this research will rely on a customised dataset that operationalises the key variables, namely OGD, accountability, and service delivery provision and outcomes. All variables have indices that can be used

---

9   Twenty-five countries are included in the study, including: Benin, Botswana, Burkina Faso, Cameroon, Côte d'Ivoire, Democratic Republic of Congo, Ethiopia, Ghana, Kenya, Malawi, Mali, Mauritius, Mozambique, Namibia, Nigeria, Rwanda, Senegal, Sierra Leone, South Africa, Swaziland, Tanzania, Togo, Uganda, Zambia, and Zimbabwe.

to compare countries across the sub-Saharan Africa region with respect to the levels of OGD and the resulting levels of accountability and service delivery. To address potential endogeneity issues, the direction of the relationship will be strengthened by the sequencing of data gathered, which would imply that OGD changes occur *prior* to changes in accountability and service delivery.

- **Open government data:** This index will draw upon the Open Data Barometer (ODB), produced on an annual basis by the World Wide Web Foundation, which scores countries on the openness of their data. In particular, the ODB analyses global trends, and provides comparative data on countries and regions using an in-depth methodology that combines contextual data, technical assessments, and secondary indicators.[10] For each year, a country is given an overall dataset score (1–100) for the completeness of OGD datasets based on a standardised rubric across a number of areas, including: maps, land ownership, census, government budget, government spending, company registration, legislation, public transport, international trade, health, education, crime statistics, environment statistics, election results, and contracts. For the purposes of the present analysis, raw scores provided by the ODB were slightly remixed to obtain a certain operationalisation of the variables based on the theoretical literature and conceptual framework.

- **Accountability:** Since the research will focus on sub-Saharan Africa, accountability data were obtained from the Mo Ibrahim Index of African Governance (IIAG), which has rated accountability in virtually every African country since 2001. The dataset relies on an accountability index based on an inclusive list of sub-indicators. As noted in the theoretical literature, there are many definitions of accountability and there are likewise many different measures of what constitutes accountability across different international institutions' indices. Accordingly, such an indexed variable – which draws on many of the principal accountability indexes including those produced by the World Bank, African Development Bank, Economist Intelligence Unit, Bertelsmann, the World Economic Forum, and others – ensures that the concept is broadly operationalised.

- **Service delivery:** Measuring service delivery can be difficult, as it has many different dimensions including access, quality, timeliness, price, efficacy, and satisfaction. For the sake of simplicity, service delivery will be operationalised in two dimensions: (i) provision of services and (ii) outcomes of services. The rationale for this is that provision (e.g. number of teachers in a district, number of books purchased) may be more easily

10 See https://opendatabarometer.org/barometer/

impacted by improvements in OGD as opposed to outcomes (e.g. primary school completion, literacy, etc.). This index will likewise draw upon the IIAG, which rates education and health at the aggregate level. Based on a review of the sub-indicators, the IIAG aggregate indicators were remixed in order to better parse out provision and outcomes in both the health and education sectors.

- **Control and intervening variables:** Based on the literature, and as highlighted in the conceptual framework above, it is evident that a number of control/intermediating variables should be included, which can affect the relationships between the key independent and dependent variables. While many additional factors can and should be considered, the statistical analysis will only operationalise the most obvious enabling factors, including the following:

  » **Publicity condition:** This index relies on the IIAG sub-index for 'Rights' which averages scores for freedom of expression, media freedom, media impartiality, censorship, freedom of association and assembly, trade unions, civil liberties, human rights, and the like.
  » **Political agency condition:** This index relies on the IIAG sub-index for 'Participation' which averages scores for political participation, freedom of political parties, CSO/NGO participation, free and fair elections, election monitoring, legitimacy of the political process, and the like.
  » **Public management condition:** This index relies on the IIAG sub-index for 'Public Management' which averages scores for public administration capacity, statistical capacity, budget management, fiscal policy, revenue collection, HR management, and the like.
  » **Client power condition:** This research proposes a customised index that relies on a variety of sub-indicators related to CSO/NGO freedom, IT literacy, mobile phone and internet penetration, media censorship, freedom of association and assembly, and access to information.
  » **Log per capita GDP:** In order to remove the income effects between countries, as well as to control for inflationary effects, this research uses log per capita GDP harmonised to 2011.

Statistical findings

Using the multi-year dataset created, each premise in the conceptual framework was tested in order to estimate associations between the principal independent variable (OGD) and the respective dependent variables (accountability and service delivery). In addition to these aggregates, key relationships were tested with respect to variables contained in the sub-indices of accountability, service provision, and service delivery outcomes, which may be of interest on their own – outside the hypothesised relationships noted in the conceptual framework. Finally, the four enabling 'conditions,' – publicity, political agency, public management, and client power – were tested in order to provide important clues regarding the necessary and sufficient conditions in each relationship.

Premise 1: Higher levels of OGD make government more accountable

To test the first premise, the aggregate accountability index is regressed on the overall average score of OGD implementation (across all dimensions of OGD). First, a positive and statistically significant correlation is found between the indexes: for a 1-point increase in OGD, accountability is expected to increase by 0.375 points in the sample of countries surveyed. Second, when controlling for the Political Agency condition index and the Publicity condition index, the association between OGD and accountability remain significant in the case of the Political Agency condition and marginally significant in the case of the Publicity condition, suggesting that all three variables are meaningful both independently and when taken together. As expected, higher levels of per capita GDP are associated with higher levels of accountability.

	(1) Accountability	(2) Accountability	(3) Accountability
OGD Average	0.375**	0.336*	0.289
	(0.156)	(0.171)	(0.180)
Political Agency		0.203**	
		(0.0770)	
Publicity			0.294***
			(0.110)
Log GDP	9.778***	7.213***	7.809***
	(1.437)	(1.420)	(1.626)
_cons	-35.19***	-26.13***	-33.41***
	(10.66)	(9.792)	(11.40)
N	86	86	86
R²	0.421	0.469	0.486

In addition to the overall accountability index, positive and significant correlations are found with respect to other accountability sub-indices, including those produced by the World Bank and the African Development Bank – thus adding to the validity of the purported relationship. In addition, greater levels of OGD are positively associated with less abuse of office (Bertelsmann) as well as less diversion of public funds (WEF). However, OGD does not seem to have a significant effect on popular corruption indexes, including Executive Corruption (VDem), Corruption in Public Officials (EIU), and Corruption in Bureaucracy (WB).

Premise 2: More accountable governments enable improved service delivery

To test the second premise, education and health service delivery – in terms of both provision and outcomes – are regressed on the accountability index. First, in the education sector, no significant effect of accountability is found on education service delivery in terms of provision or outcomes. Second, when controlling for the public management condition – which again, is the index used to measure administrative capacity – no significant association is found with respect to either education provision or outcomes. Finally, it should be noted that the coefficient for accountability becomes marginally negative and significant when controlling for public management, which is an unexpected result given that higher levels of public administration capacity would be expected to correlate with improved education outcomes.

	(1) Education Provision	(2) Education Outcomes	(3) Education Provision	(4) Education Outcomes
Accountability	0.0658	-0.140	0.162	-0.205*
	(0.115)	(0.0915)	(0.144)	(0.111)
Public Mgmt.			-0.232	0.157
			(0.194)	(0.150)
Log GDP	8.975***	17.12***	9.429***	16.82***
	(1.952)	(1.447)	(1.872)	(1.487)
_cons	-19.36	-82.99***	-15.54	-85.58***
	(11.70)	(8.614)	(12.84)	(8.452)
N	86	86	86	86
R²	0.428	0.741	0.437	0.743

However, when the same regression is performed in the health sector, different associations emerge. First, a significant correlation of accountability is found with respect to health provision and outcomes, whereby a 1-point increase in the accountability index, is correlated with an increase in the health provision index by 0.406 points and the health outcomes index by 0.167 points for the average country in the sample. In these specifications, the effect of accountability on health service provision is larger and more significant than on outcomes, since outcomes may take more time to materialise. Second, when controlling for the public management condition, accountability remains significant for health provision, suggesting that it is a driving factor in this relationship. Conversely, for health outcomes, the opposite is true: accountability loses its significance and public management shows a positive and significant effect (0.259), suggesting that public management might be a more relevant driving factor than accountability for health outcomes.

	(1) Health Provision	(2) Health Outcomes	(3) Health Provision	(4) Health Outcomes
Accountability	0.406***	0.167**	0.423**	0.0599
	(0.148)	(0.0668)	(0.174)	(0.0851)
Public Mgmt.			-0.0415	0.259*
			(0.220)	(0.132)
Log GDP	-0.271	3.196**	-0.190	2.689**
	(2.310)	(1.304)	(2.311)	(1.290)
_cons	57.37***	45.64***	58.05***	41.37***
	(12.95)	(8.727)	(14.04)	(9.044)
N	86	86	86	86
R²	0.233	0.288	0.233	0.312

Premise 3: Higher levels of OGD make governments more accountable, which subsequently enables improved service delivery

To test the third premise, service provision and outcomes in both the health and education sectors are regressed on sector-specific OGD, while studying the interaction effects of the accountability index. First, in the regression specification for education service delivery, OGD is estimated to be positively and significantly correlated with education provision and outcomes indices. Second, when controlling for the levels of accountability, the estimated coefficients for OGD on both provision and outcomes remains almost unchanged, suggesting that the link between OGD and education service delivery is independent from levels of accountability at the country level. As mentioned above, the regressions likewise find that the effect of accountability on education provision is not significant, while the effect on outcomes is marginally negative (which is an unexpected result). Finally, no interaction effects of accountability are found with respect to the relationship between OGD and education service delivery (not shown).

	(1) Education Provision	(2) Education Outcomes	(3) Education Provision	(4) Education Outcomes
OGD Education	0.201***	0.171***	0.199***	0.184***
	(0.0645)	(0.0426)	(0.0646)	(0.0420)
Accountability			0.0376	-0.167*
			(0.114)	(0.0913)
Log GDP	9.141***	15.20***	8.754***	16.92***
	(1.198)	(0.874)	(1.842)	(1.370)
_cons	-21.62**	-78.07***	-20.31*	-83.87***
	(9.568)	(7.391)	(11.02)	(8.091)
N	86	86	86	86
R²	0.487	0.759	0.488	0.775

In the regression specification for health service delivery and OGD, slightly differentiated relationships between variables emerge. First, when disaggregating between provision and outcomes, OGD is positively and significantly correlated with health outcomes but not provision. Second, when controlling for accountability, the relationship between OGD and health provision remains insignificant, and the relationship between OGD and health outcomes remains significant. This would imply that in terms of health outcomes, levels of accountability and OGD can independently have an important effect, whereas accountability – and not OGD – is an important factor for provision. Finally, the interaction effects of accountability are tested with respect to the relationship between OGD and health service delivery, and no significant effect is found (not shown).

	(1) Health Provision	(2) Health Outcomes	(3) Health Provision	(4) Health Outcomes
OGD Health	0.127	0.109**	0.0627	0.0859*
	(0.0861)	(0.0448)	(0.0667)	(0.0451)
Accountability			0.389**	0.143**
			(0.149)	(0.0699)
Log GDP	4.130***	5.068***	-0.0216	3.539***
	(1.540)	(0.993)	(2.328)	(1.326)
_cons	39.61***	36.70***	54.97***	42.36***
	(11.77)	(8.046)	(13.29)	(8.910)
N	86	86	86	86
R²	0.105	0.274	0.239	0.309

*Premise 4: Higher levels of OGD directly enable improved service delivery,
excluding the effects of accountability*

To test the final premise, service provision and outcomes are regressed against
OGD, omitting the role of accountability as suggested above with respect to
the sectoral differentiation, but testing alternative control variables such as
the Public Management condition and the Client Power condition. First, in
the education sector, OGD is estimated to have a stronger correlation with
education provision than outcomes, which may reflect the time needed for
policy to translate into results; however, it is worth highlighting that both
relationships are significant. Second, no evidence supports the claim that either
Public Management or Client Power has relevance with education provision
or outcomes. Finally, even controlling for these conditions, the relationship
between OGD and service delivery in the education sector does not change in a
meaningful way with respect to the coefficients.

	(1) Education Provision	(2) Education Outcomes	(3) Education Provision	(4) Education Outcomes	(5) Education Provision	(6) Education Outcomes
ODG Education	0.201***	0.171***	0.217***	0.181***	0.213***	0.183***
	(0.0645)	(0.0426)	(0.0661)	(0.0436)	(0.0671)	(0.0442)
Client Power			-0.130**	-0.083		
			(0.0476)	(0.0679)		
Public Mgmt.					-0.136	-0.148
					(0.169)	(0.134)
Log GDP	9.141***	15.20***	10.37***	15.99***	9.973***	16.10***
	(1.198)	(0.874)	(1.383)	(1.063)	(1.586)	(1.232)
_cons	-21.62**	-78.07***	-24.66***	-80.03***	-21.34**	-77.76***
	(9.568)	(7.391)	(10.003)	(7.491)	(9.396)	(7.632)
N	86	86	86	86	86	86
R^2	0.487	0.759	0.505	0.764	0.493	0.764

Drilling down deeper, strong associations are found between OGD and certain
aspects of education service delivery. As mentioned, when education service
delivery is regressed on the OGD education index, a positive and significant
association is found in terms of provision. Importantly, the positive and
significant correlation between education provision and OGD is strongly driven
by indicators linked to quality and management of education provision. In terms
of outcomes, when education service delivery outcomes are regressed on the
OGD education index, a positive and significant association is found in terms
of certain outcomes as well. For instance, the association between OGD and
education completion is only significant for primary levels of education, whereas

it is insignificant for secondary and tertiary completion.

In the health sector, OGD displays associations which are heterogeneous. First, OGD is found to be positively associated with health outcomes, but there is not such statistical evidence of an association with health provision. Second, with respect to health provision, Public Management capacity appears to be a more important determining factor than OGD (see column 5). Likewise, when controlling for the public management condition, the OGD coefficient linked to health outcomes remains significant, but slightly declines, suggesting that the impact of OGD on health outcomes (see column 6) is partly driven by changes in the index of Public Management, although both are independently significant. Finally, when controlling for the client power condition, this relationship remains unaltered, suggesting that this condition has little impact on the relationship between OGD and health service delivery in the countries included. In summary, it can be shown that overall, public management plays a role in health provision and outcomes, while OGD seems to further contribute to the outcome channel but not provision.

	(1) Health Provision	(2) Health Outcomes	(3) Health Provision	(4) Health Outcomes	(5) Health Provision	(6) Health Outcomes
OGD Health	0.127	0.109**	0.136	0.105**	0.0877	0.0809*
	(0.0861)	(0.0448)	(0.0852)	(0.0465)	(0.0762)	(0.0447)
Client Power			-0.0960	0.0412		
			(0.0660)	(0.0049)		
Public Mgmt.					0.394*	0.289***
					(0.229)	(0.109)
Log GDP	4.130***	5.068***	5.079***	4.661***	1.603	3.212**
	(1.540)	(0.993)	(1.785)	(1.059)	(2.242)	(1.220)
_cons	39.61***	36.70***	37.08***	37.79***	39.91***	36.92***
	(11.77)	(8.046)	(12.408)	(8.075)	(11.96)	(7.782)
N	86	86	86	86	86	86
R^2	0.105	0.274	0.116	0.278	0.155	0.327

At a disaggregated level, a strong association is found between OGD and certain aspects of health service delivery. As seen in the previous table, OGD and health provision are not significantly correlated; however, there is a significant association between OGD and sub-indexes related to immunisations, including DPT and Hepatitis B. No correlations exist for other health provision indicators including measles immunisation and provision of anti-retroviral treatments. With respect to specific outcome indictors, it should be noted that the OGD index is significantly correlated with the index of health outcomes, as well as with child mortality outcomes and WHO disease prevention statistics.

Conclusions

Principles and generalisations inferred from the results

With respect to the four original premises put forth above, and the various regression analyses carried out, a number of generalisations can be drawn with respect the effects of OGD. Importantly, while certain relationships may hold in each of the premises analysed, when taken together, larger conclusions may be drawn with respect to relationships between variables. Accordingly, these initial conclusions can help to bolster the empirical evidence base for the various public sector theories advocated in the literature as well as to highlight areas for future research on the topic.

Premise 1: Higher levels of OGD make government more accountable

With respect to the first premise, this analysis confirms a strong positive association between OGD and accountability in the countries surveyed over the period. When controlling for other contextual factors such as the levels of political freedoms, civil rights, and access to information (e.g. publicity condition) as well as political accountability, including the presence of formal electoral mechanisms (e.g. political agency condition), this analysis likewise finds these factors significant, independent of OGD. This suggests that *all three factors* can be considered important elements in improving government accountability, in line with the premised theory. ***As such, while OGD is necessary for improved accountability, it is not sufficient, as publicity and political agency must also be included – thus confirming the earlier theoretical and empirical contributions.*** Even when using other accountability indexes – for instance, related accountability indices prepared by the World Bank and African Development Bank – these relationships hold, thus reinforcing the validity of this claim.

Figure 2. Effects of OGD on accountability

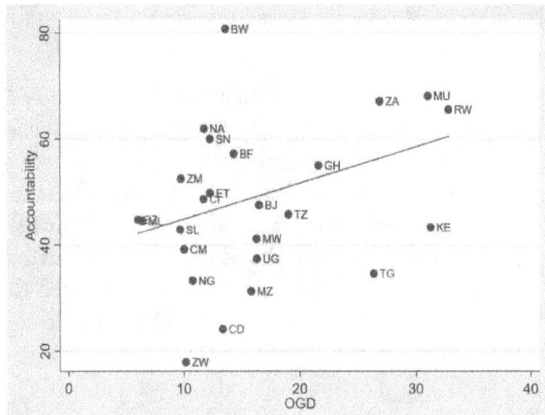

Premise 2: More accountable governments enable improved service delivery

With respect to the second premise, this analysis provides evidence that accountability – in particular, those governments exhibiting higher levels of accountability – correlates with improved service delivery; however, there are differentiated effects across education provision and outcomes and health provision and outcomes. This is an important finding in and of itself, as it highlights the fact that sectoral specificities between the health and education sectors – in terms of how decisions are made, how services are rendered, and how clients consume these services – have a strong determining effect. Another important specification highlighted by the research is the importance of differentiating between service provision, which may be more readily governed by public policy decisions and implementation, versus service outcomes, which may be influenced by a myriad of other contextual factors and take years to materialise.

To be sure, in the education sector, this analysis finds no correlation between levels of accountability and education provision and outcomes. Even when controlling for levels of public sector management capacity, which is an important tool for governments to translate their accountability structures and practices into improvements in service delivery, this initial analysis finds no association. Returning to the literature, there can be a number of important explanations for this seemingly counterintuitive finding. For instance, in the education sector in particular, there are a number of sector-level political economy factors which may confound the impacts of accountability and public management capacity. These include, among others: (i) the potential for excludability, which affects the power relationships between those who demand access to education (e.g. students, parents, etc.) and those who control access to those services (e.g. school administrators); (ii) principal-agent asymmetries, whereby parents, as agents, make decisions for their children, who are the 'rightful' principals of the service; (iii) the fact that education is a 'merit' good, whereby its consumption is compulsory for most clients (e.g. students themselves); (iv) education outcomes often require 'co-production,' which means that frontline providers and recipients (e.g. students and parents) both need to invest their time in achieving outcomes; and (v) issues related to information asymmetries and visibility, whereby it is difficult to measure the quality of education inputs (e.g. curriculum and instructional quality), which can have knock-on effects with respect to the political salience of policy decisions. All of these – and other – sectoral level dynamics may explain why a positive association with accountability or public management in the education sector is not apparent in the sample selection of countries and years studied. However, this finding should not be interpreted to suggest that government accountability is not an important determining factor in the delivery of education more broadly, as accountable institutions remain *sine qua non* for ensuring efficacy and efficiency of the public sector.

Conversely, in the health sector, public accountability appears to have a significant association with service provision and outcomes. However, when controlling for levels of public sector management capacity, which is an equally important determinant in how services are delivered, the results become more nuanced. In particular, it would appear that accountability is a more important condition for health provision, whereas public management capacity seems to absorb the effect that accountability has with respect to health outcomes. These differentiated effects between health provision and outcomes may likewise have some grounding in the literature related to sector-level political economy. For instance, a number of characteristics make health service delivery significantly more 'murky' than other sectors, including, among others: (i) the difference between preventative and curative services, the latter of which may dominate, as many health services are consumed on an 'emergency' basis and do not involve continual monitoring; (ii) the presence of information asymmetries whereby patients do not have the same knowledge of the quality of service inputs as healthcare providers; (iii) the high level of individual discretion in providing health services, whereby providers are required to make complex decisions, which are difficult to monitor; and (iv) the professional dominance of healthcare providers, who are able to better politically organise in ways that allow them to resist strict government control and regulation. These – and other – sectoral characteristics may go a long way in providing explanations as to why health provision is more associated with public management capacity and accountability, whereas health outcomes have a strong linkage with public management but not accountability. While the intricacies of these relationships are outside the scope of this paper, these findings indicate a need for future qualitative political economy analysis at the sector level.

Premise 3: Higher levels of OGD make governments more accountable, which subsequently enables improved service delivery

With respect to the third premise, this analysis provides evidence that accountability can play an important role in translating OGD into improved service delivery, but again, there are important distinctions between education provision and outcomes and health provision and outcomes. As mentioned above, the principal-agent relationships of accountability in the education sector may differ significantly from those in the health sector, especially where responsibilities are delegated to head of household as opposed to the eventual consumers of these services. Likewise, the nature of service delivery can have important determinants regarding how accountability affects provision and outcomes, given, for instance, differences between how access, quality, and oversight may translate into achievements in the education sector and how preventative and curative services may alter monitoring, oversight, and compliance in the health sector.

In the education sector, the findings suggest that OGD is positively and significantly correlated with both provision and outcomes. Importantly, these relationships maintain their significance when controlling for accountability, suggesting that the link between OGD and education service delivery is independent from levels of accountability at the country level. Returning to the finding noted above, accountability does not appear to be a significant condition in intermediating the relationship between OGD and education service delivery, which suggests a direct effect of OGD on provision and outcomes. Concretely, this means that the present analysis finds no evidence of the 'long-route' to service delivery in the education sector, whereby OGD fosters increased accountability, which then leads to improved service delivery. Rather, this may mean that policymakers can use OGD simply as a management tool – outside a democratic accountability process – to make top-down policy decisions that affect service delivery in education in the countries concerned. Alternatively, this may mean that citizens are able to directly influence service delivery from frontline providers using social (non-formal) accountability mechanisms, including collective engagement through parent-teacher committees, participatory budgeting and oversight exercises, or direct citizen pressure on service providers such as petitions and protests. Both of these alternate 'short-route' hypotheses are tested subsequently under the fourth premise.

Figure 3. Effects of OGD on education provision and outcomes

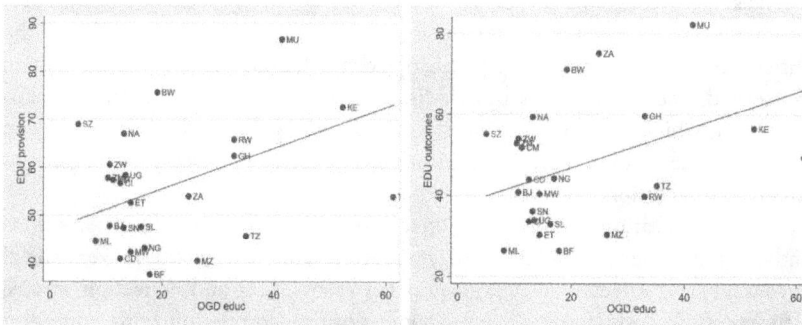

Conversely, in the health sector, OGD is not associated with improved service provision, but it is significantly correlated with improved outcomes. When controlling for the effects of accountability on this relationship, the analysis finds that it has an independent relationship with respect to service provision but works in tandem with OGD with respect to health outcomes. This would therefore suggest that there may be evidence of the 'long-route' to service delivery in the health sector, whereby OGD fosters increased accountability, which then leads to improved service delivery – with the caveat that 'service delivery' is conceptualised on the basis of outcomes, not provision. Again, the reasons for these differentiated effects are likely vast and varied and would likewise require a level of qualitative analysis outside of this initial statistical estimation. Future research on this topic should thus work to parse out the sector-level details to provide more adequate explanations supporting these associations.

Figure 4. Effects of OGD on health provision and outcomes

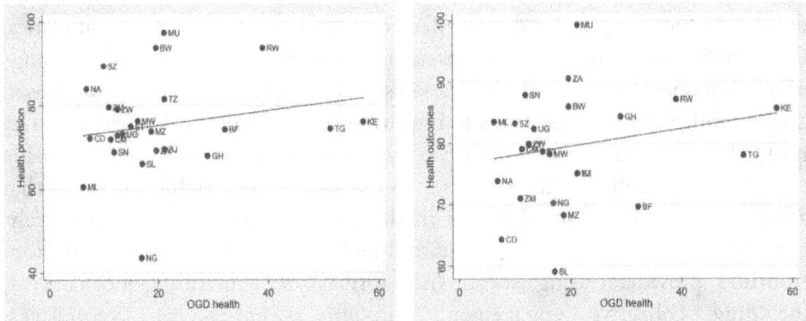

Premise 4: Higher levels of OGD directly enable improved service delivery, excluding the effects of accountability

With respect to the fourth premise, this analysis finds differentiated effects of public sector management capacity and client power (e.g. social accountability) mechanisms. Again, these dynamics may be more largely related to the differences in how services are planned, executed, and evaluated in the sectors concerned. These differences are also linked to particular aspects with respect to provision, which may be more easily controlled by public policy and execution than outcomes, which represent the long-term effects of service delivery policies and implementation.

In the education sector, additional testing of other enabling conditions indicates that neither public management capacity nor client power (social accountability) affects the relationship between OGD and education service delivery. OGD is found to have a more positive correlation with education provision than outcomes (although both are significant), which may reflect the time needed for policy to translate into results. For their part, client power and public management capacity appear to have little influence on these results, suggesting that some key explanatory gaps remain in the education sector, which may have several interpretations. First, in the absence of any additional control or intervening variables, it may suggest a direct relationship between OGD and education service delivery, meaning that OGD is an important determinant in the education sector. Conversely, it may suggest other lurking variables, since education provision and outcomes are likely to be intimately intertwined with other contextual factors such as family income, access disparities between urban

226

and rural areas, as well as other cross-sectoral issues such as infrastructure, nutrition, as well as conflict and gender dynamics. Finally, it may suggest certain political economy dynamics, including rent seeking, clientelism, patronage, and other forms of policy capture, which, as noted above, can distort principal-agent relationships between citizens and public officials. Again, while the present analysis does not attempt to answer these questions, this suggests the need for additional future quantitative and qualitative research that delves deeper into sectoral level service delivery dynamics.

As noted above, in the health sector, the findings suggest a positive association between OGD and health outcomes, but not provision. Importantly, the analysis finds that part of the impact of OGD on health indicators is driven by changes in the index of public management, although not client power, which appears to have a direct and important impact on both health provision and outcomes. Accordingly, the findings point to the role that public management can play as a predictor of better health provision and outcomes, while such a relationship is not apparent in education (as was also demonstrated in Premise 2). This finding must also be taken in context with previous findings on the importance of accountability, which likewise has differentiated associations between health provision and outcomes. On the aggregate, these findings suggest that accountability and public management capacity – but not OGD – impact health provision, whereas OGD and public sector management capacity – but not accountability – impact health outcomes. Finally, as in education, client power does not seem to play a role in health provision and outcomes, which is a counterintuitive finding. Thus, future research may need to better conceptualise this variable or employ methods to better specify the effects of client power.

Limitations

Although this research has considerable promise, there are a number of limitations which should be acknowledged, including, most importantly, that this analysis only claims to be an *initial* assessment of the relationship between the variables. While it would be ideal to have more years of data – and more observations included in the dataset – the analysis uses data that are *currently* available, as OGD only started to be measured in 2012/2013. In this regard, OLS regressions are instructive in teasing out initial associations between variables; however, as more data become available, increasingly more sophisticated estimations can be performed in the future. For instance, it would be ideal to include a country fixed-effects estimation as it would demonstrate that the results are not driven by omitted variable bias, which are country-specific and do not change over time. Unfortunately, including country fixed-effects estimation at this point in time would lead to a significant loss of degrees of freedom to estimate the parameters of interest. As such, there is an inherent trade off between unbiasedness/consistency that would be provided by the fixed-effects estimation versus being able to detect parameters using this initial OLS estimation.

Second, given the fact that OGD is a rather recent phenomenon, it is difficult to suggest a purely *causal* relationship between variables. While a lag of nine months has been introduced between the availability of OGD statistics and changes in accountability and service provision, it is likely that these changes may take more time to materialise. Accordingly, the present analysis only purports to provide evidence of associations between variables, to help support or refute existing theoretical propositions. Relatedly, challenges exist with certain indicators and it is necessary to ensure that there are no significant issues of endogeneity and simultaneous causality. For instance, more accountable governments *may* be more inclined to support an OGD initiative in the first place, which would reverse the causal relationship. As such, it is difficult to argue at this early stage whether OGD precedes accountability or accountability precedes OGD. Further quantitative and qualitative methods will need to be employed (e.g. process tracing, case study analysis, and qualitative comparative analysis) to establish a higher degree of causal certainty in the future.

Finally, there may be difficulties extrapolating the relationships demonstrated in the sub-Saharan context to other countries, particularly upper-middle-income and high-income countries. To be sure, the analysis finds the coefficient relating OGD and accountability to be of a significant magnitude, which is due primarily to the low levels of OGD prevalent in the sub-Saharan African region. For this reason, one cannot expect the same gains from OGD in countries with already high levels of accountably and service delivery as would be possible in country contexts with relatively low starting points. This is precisely the reason why the present study omits other regions and focuses on a sample of countries in sub-Saharan Africa, a place where OGD has the greatest potential for improving accountability and service delivery.

Recommendations and future research

Given that this paper seeks to present an initial assessment of OGD, account-ability, service delivery, as well as some of the other relevant 'conditions' as gleaned from the theoretical literature, it is perhaps too early to put forth any substantive recommendations on what needs to be done to better enable OGD to reach its full potential. In its place, the initial findings presented here suggest a number of streams of future research that need to be pursued, including, inter alia: (i) deeper sectoral level analysis in education and health to better understand key provision and outcome dynamics as well as their linkages to accountability relationships; (ii) more robust understanding of relationships that do not manifest themselves statistically – such as the client power condition and related social accountability mechanisms – which ostensibly have a key role in the impact of OGD; (iii) deeper country-level analysis using qualitative methods to establish typologies of country behaviour, which can help to provide key explanations for the performance of OGD, accountability, and service delivery across different contexts; (iv) more research and recommendations for sub-national OGD, which is where most frontline service delivery takes place (e.g. hospitals, clinics, schools, etc.); (v) more granular analysis of the end-users of OGD as well as the outputs of OGD use (e.g. dashboards, visualisations, journalism, etc.); and finally, as more years of data become available, (vi) the introduction of more robust statistical methods (e.g. fixed-effects models) as well as more sophisticated qualitative methods (e.g. process tracing and qualitative comparative analysis) in order to strengthen the causal links between OGD, accountability, and service delivery.

Annexures to Chapter 10

Annex. 1: Regression Specifications

Premise 1: Higher levels of open data availability make government more accountable

$$Accountability_i = \alpha + \beta_1\, OGD\, Average_i + \beta_2 \log GDP_i + \beta_3\, Political\, Agency_i + \beta_4\, Publicity_i + \varepsilon_i$$

Premise 2: More accountable governments enable improved service delivery

$$Edu.Provisions_i = \alpha + \beta_1\, Accountability_i + \beta_2 \log GDP_i + \beta_3\, Public\, management_i + \varepsilon_i$$
$$Edu.Outcomes_i = \alpha + \beta_1\, Accountability_i + \beta_2 \log GDP_i + \beta_3\, Public\, management_i + \varepsilon_i$$
$$Health\, Provisions_i = \alpha + \beta_1\, Accountability_i + \beta_2 \log GDP_i + \beta_3\, Public\, management_i + \varepsilon_i$$
$$Health\, Outcomes_i = \alpha + \beta_1\, Accountability_i + \beta_2 \log GDP_i + \beta_3\, Public\, management_i + \varepsilon_i$$

Premise 3: Higher levels of open data availability enables improved service delivery, taking into account the effects of political accountability

$$Edu.Provisions_i = \alpha + \beta_1\, OGD\, education_i + \beta_2 \log GDP_i + \beta_3\, Accountability_i + \beta_4\, OGD\, educ_i * accountability_i + \varepsilon_i$$
$$Edu.Outcomes_i = \alpha + \beta_1\, OGD\, education_i + \beta_2 \log GDP_i + \beta_3\, Accountability_i + \beta_4\, OGD\, educ_i * accountability_i + \varepsilon_i$$
$$Health\, Provisions_i = \alpha + \beta_1\, OGD\, health_i + \beta_2 \log GDP_i + \beta_3\, Accountability_i + \beta_4\, OGD\, health_i * accountability_i + \varepsilon_i$$
$$Health\, Outcomes_i = \alpha + \beta_1\, OGD\, health_i + \beta_2 \log GDP_i + \beta_3\, Accountability_i + \beta_4\, OGD\, health_i * accountability_i + \varepsilon_i$$

Premise 4: Higher levels of open data availability enable improved service delivery, excluding the effects of political accountability

$$Edu.Provisions_i = \alpha + \beta_1\, OGD\, education_i + \beta_2 \log GDP_i + \beta_3\, Client\, Power_i + \beta_4\, Public\, Management_i + \varepsilon_i$$
$$Edu.Outcomes_i = \alpha + \beta_1\, OGD\, education_i + \beta_2 \log GDP_i + \beta_3\, Client\, Power_i + \beta_4\, Public\, Management_i + \varepsilon_i$$
$$Health\, Provisions_i = \alpha + \beta_1\, OGD\, health_i + \beta_2 \log GDP_i + \beta_3\, Client\, Power_i + \beta_4\, Public\, Management_i + \varepsilon_i$$
$$Health\, Outcomes_i = \alpha + \beta_1\, OGD\, health_i + \beta_2 \log GDP_i + \beta_3\, Client\, Power_i + \beta_4\, Public\, Management_i + \varepsilon_i$$

Annex. 2: Indexed Indicators

Accountability Index (including indicator sources)

Adjusted Accountability Index

- Accountability, Transparency & Corruption in the Public Sector (AfDB/WB)
- Accountability, Transparency & Corruption in the Public Sector (AfDB)
- Accountability, Transparency & Corruption in the Public Sector (WB)
- Corruption & Bureaucracy (WB)
- Corruption in Government & Public Officials (EIU)
- Diversion of Public Funds (WEF)
- Accountability of Public Officials (EIU)
- Public Sector Corruption Investigation (GI)
- Public Sector Corruption Bodies (GI)
- Corruption Investigation (GI)
- Prosecution of Abuse of Office (BS)
- Accountability, Transparency & Corruption in Rural Areas (IFAD)

Sub-Indicators removed for endogeneity

- Access to Information (GI)
- Public Information (GI)
- Access to Legislative Processes & Documents (GI)

Source: Mo Ibrahim Index of African Governance Indicators 2017

Service Delivery Indices (including indicator sources)

Education Service Delivery	Health Service Delivery
Overall Education Index (provision + outcomes)	Overall Health Index (provision + outcomes)
Education Provision Sub-Index • Education Provision & Quality (BS) • Education System Quality (WEF) • Ratio of Pupils to Teachers in Primary School (UNESCO)	Health Provision Sub-Index • Basic Health Services (AFR) • Public Health Campaigns (GI) • Immunisation (WB/WHO) • Immunisation against Measles (WB) • Immunisation against DPT (WB) • Immunisation against Hepatitis B (WHO) • Antiretroviral Treatment (ART)
Education Outcomes Sub-Index • Primary School Completion (WB) • Secondary School Enrolment (UNESCO) • Tertiary Education Enrolment (UNESCO) • Literacy (UNESCO)	• Provision (UNAIDS) • ART Provision (UNAIDS) • ART Provision for Pregnant Women (UNAIDS) • Health Outcomes Sub-Index • Child Mortality (IGME) • Maternal Mortality (MMEIG) • Undernourishment (WB) • Disease (WHO) • Malaria (WHO) • Tuberculosis (WHO)

Source: Mo Ibrahim Index of African Governance Indicators 2017

Conditional Indicators

Publicity Condition	Political Agency Condition
• Freedom of Expression (BS/RSF/V-Dem/GI)	• Political Participation (EIU/FH/V-Dem)
• Freedom of Expression (BS)	• Political Participation (EIU)
• Media Freedom (RSF)	• Political Pluralism (FH)
• Media Impartiality (V-Dem)	• Freedom of Political Parties (V-Dem)
• Freedom of Expression (V-Dem)	• Civil Society Participation (BS/V-Dem/GI)
• Censorship (GI)	• Civil Society Participation (BS)
• Media Censorship (GI)	• Civil Society Inclusion (V-Dem)
• Online Censorship (GI)	• Freedom of NGOs (GI)
• Freedom of Association & Assembly (BS/GI)	• Barriers to NGO Operations (GI)
• Freedom of Association & Assembly (BS)	• Persecution of NGOs (GI)
• Freedom of Association (GI)	• Harassment of NGOs (GI)
• Freedom of Association (GI)	• Free & Fair Elections (BS/CDD/V-Dem)
• Trade Unions (GI)	• Free & Fair Elections (BS)
• Civil Liberties (BS/FH)	• Free & Fair Executive Elections (CDD)
• Protection of Civil Liberties (BS)	• Free & Fair Elections (V-Dem)
• Civil Liberties (FH)	• Election Monitoring Agencies (V-Dem/GI)
• Human Rights Conventions (UNOLA/OHCHR)	• Election Management Bodies (V-Dem)
• Human Rights Violations (EIU)	• Election Monitoring Agencies (GI)
• Protection against Discrimination (GI)	• Election Monitoring Agency Independence (GI)
• Protection against Ethnic Discrimination (GI)	• Election Monitoring Agency Reporting (GI)
• Protection against Religious Discrimination (GI)	• Legitimacy of Political Process (BS)
Public Management Condition	**Client Power Condition**
• Governmental Statistical Capacity (WB)	• Access to Information (GI)
• Civil registration (GI)	• Access to Public Information (GI)
• Birth Registration (GI)	• Access to Legislative Information (GI)
• Death Registration (GI)	• Online Public Services (UNDESA)
• Public Administration (AfDB/WB)	• Social Unrest (EIU/ACLED)
• Public Administration (AfDB)	• Social Unrest (EIU)
• Public Administration (WB)	• Riots & Protests (ACLED)
• Diversification (AfDB/OECD/UNDP)	• Civil Society Participation (BS/V-Dem/GI)
• Budget Management (AfDB/WB)	• Civil Society Participation (BS)
• Budget Management (AfDB)	• Civil Society Inclusion (V-Dem)
• Budget Management (WB)	• Freedom of NGOs (GI)
• Budget Balance (AfDB/AUC/UNECA)	• Barriers to NGO Operations (GI)
• Fiscal Policy (AfDB/WB)	• Persecution of NGOs (GI)
• Fiscal Policy (AfDB)	• Harassment of NGOs (GI)
• Fiscal Policy (WB)	• Freedom of Expression (BS)
• Revenue Mobilisation (ICTD/AfDB/WB)	• Media Freedom (RSF)
• Taxation Capacity (ICTD)	• Media Impartiality (V-Dem)
• Revenue Collection (AfDB/WB)	• Freedom of Expression (V-Dem)
• Revenue Collection (AfDB)	• Censorship (GI)
• Revenue Collection (WB)	• Media Censorship (GI)
• Transparency of State-owned Companies (GI)	• Online Censorship (GI)
	• Freedom of Association & Assembly (BS/GI)
	• Freedom of Association & Assembly (BS)
	• Freedom of Association (GI)
	• Freedom of Association (GI)
	• Trade Unions (GI)
	• Digital & IT Infrastructure (EIU/ITU)
	• IT Infrastructure (EIU)
	• Mobile Phone Subscribers (ITU)
	• Household Computers (ITU)

Source: Mo Ibrahim Index of African Governance Indicators 2017

Annex. 3: Summary Statistics

	Mean	Median	Standard Deviation	Minimum	Maximum
OGD Overall Score	16.6	13.6	8.8	3.3	45.7
OGD Education	20.8	12.5	15.8	3.8	77.5
OGD Health	20.0	12.5	15.1	3.8	77.5
Accountability Index Score	49.4	47.9	14.6	12.3	83.4
Education Provision Index Score	55.8	56.3	12.6	36.7	88.4
Education Outcomes Index Score	47.3	43.3	15.5	24.9	82.9
Health Provision Index Score	75.2	74.2	12.0	35.6	98.5
Health Outcomes Index Score	79.5	79.7	8.6	58.9	99.4
Political Agency Index Score	59.9	61.3	19.3	15.1	89.9
Publicity Index Score	52.5	52.8	14.3	23.5	80.6
Client Power Index Score	54.7	55.4	15.6	19.7	82.7
Public Management Index Score	52.7	50.3	8.8	34.2	74.7
N	=86				

Country Observations by Year (n=86)

	2013	2014	2015	2016
Benin	X	X	X	X
Botswana	X	X	X	X
Burkina Faso	X	X	X	X
Cameroon	X	X	X	X
Côte d'Ivoire				X
Democratic Republic of Congo				X
Ethiopia	X	X	X	X
Ghana	X	X	X	X
Kenya	X	X	X	X
Malawi	X	X	X	X
Mali	X	X	X	X
Mauritius	X	X	X	X
Mozambique		X	X	X
Namibia	X	X	X	X
Nigeria	X	X	X	X
Rwanda	X	X	X	X
Senegal	X	X	X	X
Sierra Leone		X	X	X
South Africa	X	X	X	X
Swaziland			X	
Tanzania	X	X	X	X
Togo				X
Uganda	X	X	X	X
Zambia	X	X	X	X
Zimbabwe	X	X	X	X

233

REFERENCES

Arrow KJ (1951) *Social Choice and Individual Values*. USA: Yale University Press

Atz U, Heath T & Fawcett J (2015) *Benchmarking open data automatically*. Technical Report 000. UK: Open Data Institute

Awortwi N & Nuvunga A (2019) Sound of one hand clapping: Information disclosure for social and political action for accountability in extractive governance in Mozambique. *IDS Working Paper* No. 523. Brighton: IDS

Barr A, Bategeka L, Guloba M, Kasirye I, Mugisha F, Serneels P & Zeitlin A (2012) Management and motivation in Ugandan primary schools: An impact evaluation report. *PEP Working Papers*. DOI: 10.22004/ag.econ.164412

Björkman M & Svensson J (2009) Power to the people: Evidence from a randomized field experiment on community-based monitoring in Uganda. *The Quarterly Journal of Economics* 124(2): 735–769

Black D (1958) *The Theory of Committees and Elections*. Cambridge: Cambridge University Press

Borowiak CT (2011) *Accountability and Democracy: The Pitfalls and Promise of Popular Control*. Oxford: Oxford University Press

Buchanan JM & Tullock G (1962) *The Calculus of Consent: Logical Foundations of Constitutional Democracy*. Ann Arbor, Michigan: University of Michigan Press

Carothers T & Brechenmacher S (2014) *Accountability, transparency, participation, and inclusion: A new development consensus?* Washington, DC: Carnegie Endowment for International Peace

Carruthers CK & Wanamaker MH (2015) Municipal housekeeping: The impact of women's suffrage on public education. *Journal of Human Resources* 50(4): 837–872

Chuhan-Pole P (2014) Country policy and institutional assessment (CPIA) Africa: Assessing Africa's policies and institutions. Washington, DC: World Bank Group

Dahl RA (1999) Can international organizations be democratic? A skeptic's view. In: I Shapiro & C Hacker-Cordón (eds) *Democracy's Edges*. Cambridge: Cambridge University Press. pp. 19–36

Dahl RA & Lindblom CE (1976) *Politics, Economics, and Welfare*. Chicago: University of Chicago Press

Davies T (2014) *Open data in developing countries – Emerging insights from Phase I (2)*. Berlin: World Wide Web Foundation. http://www.opendataresearch.org/sites/default/files/publications/Phase 1 – Synthesis – Full Report-print.pdf

De Tocqueville A (1956) *Democracy in America* Abridged Edn. R Heffner (ed.). New York: New American Library

Devarajan S & Khemani S (2016) If politics is the problem, how can external actors be part of the solution? (English). *Policy Research Working Paper* No. WPS 7761. Washington, DC: World Bank Group

Downs A (1957) *An Economic Theory of Democracy*. New York: Harper

Duflo E, Dupas P & Kremer M (2012) School governance, teacher incentives, and pupil-teacher ratios: Experimental evidence from Kenyan primary schools. *NBER Working Paper* No. 17939. Cambridge, MA: National Bureau of Economic Research

Farhan H, D'Agostino D & Worthington H (2012) *Web Index 2012*. World Wide Web Foundation

Ferraz C & Finan F (2008) Exposing corrupt politicians: The effects of Brazil's publicly released audits on electoral outcomes. *The Quarterly Journal of Economics* 123(2): 703–745

Fox J (2007) The uncertain relationship between transparency and accountability. *Development in Practice* 17(4-5): 663–671

Fox J (2014) Social accountability: What does the evidence really say? *GPSA Working Paper* No. 1. Washington, DC: World Bank Group

Fujiwara T (2015) Voting technology, political responsiveness, and infant health: Evidence from Brazil. *Econometrica* 83(2): 423–464

Gaventa J & McGee R (2013) The impact of transparency and accountability initiatives. *Development Policy Review* 31(1): s3–s28

Global OGD Initiative (2015) *Global OGD Initiative Survey and Interview Report* www.opendatainitiative.org/survey-and-interview-report/

Goldfrank B (2006) Lessons from Latin American experience in participatory budgeting. Presentation at the Latin American Studies Association Meeting, San Juan, Puerto Rico, March 2006

Grant RW & Keohane RO (2005) Accountability and abuses of power in world politics. *American Political Science Review* 99(1): 29–43

Gruening G (2001) Origin and theoretical basis of new public management. *International Public Management Journal* 4(1): 1–25

Gurin J (2014) *Open Data Now: The Secret to Hot Startups, Smart Investing, Savvy Marketing, and Fast Innovation.* New York: McGraw Hill Professional

Hazell R, Worthy B & Glover M (2010) The Impact of the Freedom of Information Act on Central Government in the UK: Does FOI work? UK: Palgrave Macmillan

Herzog T & Holm J (2015) Financing investments in data literacy and use: Issues paper for data revolution consultation. Washington, DC: World Bank and NASA

Hood C (1991) A public management for all seasons? *Public Administration* 69(1): 3–19

Joshi A (2013) Background paper on service delivery: Review of impact and effectiveness of transparency and accountability initiatives. Brighton, UK: Institute of Development Studies

Kendall C, Nannicini T & Trebbi F (2015) How do voters respond to information? Evidence from a randomized campaign. *American Economic Review* 105(1): 322–353

Keohane RO (2003) Global Governance and Democratic Accountability. In: D Held & M Koenig-Archibugi (eds) *Taming Globalization: Frontiers of Governance.* Oxford: Blackwell. pp. 130–159

Khemani S, Dal BE, Ferraz C, Finan FS, Stephenson Johnson CL, Odugbemi AM, Thapa D & Abrahams SD (2016) *Making politics work for development: Harnessing transparency and citizen engagement.* Policy Research Reports. Washington, DC: World Bank Group

Kosack S & Fung A (2014) Does transparency improve governance? *Annual Review of Political Science* 17: 65–87

Kudamatsu M (2012) Has democratization reduced infant mortality in sub-Saharan Africa? Evidence from micro data. *Journal of the European Economic Association* 10(6): 1294–1317

Larreguy HA, Marshall J & Snyder JM (2014) Revealing malfeasance: How local media facilitates electoral sanctioning of mayors in Mexico. *NBER Working Paper* No. 20697. Michigan: National Bureau of Economic Research

Manyika J, Chui M, Groves P, Farrell D, Van Kuiken S & Almasi Doshi E (2013) *OGD: Unlocking innovation and performance with liquid information.* Washington, DC & New Jersey: McKinsey Global Institute

Misra V & Ramansakar P (2007) Case study 1: Andhra Pradesh, India: Improving health services through community score cards: Learning notes. *Social Accountability Series.* Washington, DC: World Bank Group

Mungiu-Pippidi A (2014) *Quantitative report on causes of performance and stagnation in the global fight against corruption.* ANTICORRP Project https://www.againstcorruption. eu/wp-content/uploads/2015/12/MS2-Quantitative-report-on-causes-of-performance-and-stagnation-in-the-global-fight-against-corruption1.pdf

Myles J & Quagagno J (2009) Political theories of the welfare state. *Social Service Review* 76(1): 34–57

Nyqvist MB, De Walque D & Svensson J (2014) Information is power: Experimental evidence on the long-run impact of community-based monitoring. *Policy Research Working Paper* No. 7015. Washington, DC: World Bank Group

Olson M (1965) *The Logic of Collective Action: Public Action and the Theory of Groups.* Cambridge, MA: Harvard University Press

Open Democracy Advice Centre (2003) Turning the right to information law into a living reality: Access to information and the imperative of effective implementation. Cape Town: Open Democracy Advice Centre. https://www.humanrightsinitiative.org/programs/ai/rti/international/laws_papers/southafrica/Calland%20-%20Turning%20FOI%20law%20into%20living%20reality%20-%20Jan-03.pdf

Open Data Charter (2015) International open data charter. https://opendatacharter.net/wp-content/uploads/2015/10/opendatacharter-charter_F.pdf

Open Data Institute (2015a) Supporting sustainable development with open data. White Paper 000. UK: Open Data Institute

Open Data Institute (2015b) Open data in government: How to bring about change. White Paper 001. UK: Open Data Institute

Open Knowledge Foundation (2012, 12 May) Open data handbook. http://opendefinition.org/od/index.html

Open Society Justice Initiative (2006) *Transparency and silence: A survey of access to information laws and practices in 14 countries.* New York: Open Society Institute

Ostrom E (2005) *Understanding Institutional Diversity.* Princeton, NJ: Princeton University Press

Pandey P, Goyal S & Sundararaman V (2008a) Community participation in public schools: The impact of information campaigns in three Indian states. *Policy Research Working Paper* No. 4776. Washington, DC: World Bank Group

Pandey P, Goyal S & Sundararaman V (2008b) Public participation, teacher accountability, and school outcomes: Findings from baseline surveys in three Indian states. *Policy Research Working Paper* No. 4777. Washington, DC: World Bank Group

Peisakhin L & Pinto P (2010) Is transparency an effective anti-corruption strategy? Evidence from a field experiment in India. *Regulation & Governance* 4(3): 261–280

Peixoto T (2013) The uncertain relationship between open data and accountability: A response to Yu and Robinson's 'The New Ambiguity of Open Government.' *UCLA Law Review Discourse* 60: 200–248

Ravindra A (2004) An assessment of the impact of Bangalore citizen report cards on the performance of public agencies. *Evaluation Capacity Development Working Paper* No. 12. Washington, DC: World Bank, Operations Evaluation Department

Reinikka R & Svensson J (2004) The power of information: Evidence from a newspaper campaign to reduce capture. *World Bank Policy Research Working Paper* No. 3239. https://papers.ssrn.com/sol3/papers.cfm?abstract_id=610280

Reinikka R & Svensson J (2011) The power of information in public services: Evidence from education in Uganda. *Journal of Public Economics* 95(7-8): 956–966

Samuelson PA (1954) The pure theory of public expenditure. *Review of Economics and Statistics* 36(4): 387–389

Shakespeare S (2013) *Shakespeare Review: An independent review of public sector information.* London: BIS

Stott A (2014) OGD for economic growth: Transport and ICT global practice. Washington, DC: World Bank

Taylor FW (1911) *The Principles of Scientific Management.* New York & London: Harper & Brothers

Tinholt D (2013) *Characterization Study of the Infomediary Sector.* Spanish Open Data Portal Annual Report. Paris: Capgemini

Tisné M (2010) Transparency, participation and accountability: Definitions. Unpublished Background Note for Transparency and Accountability Initiative, Institute of Development Studies

Trapnell S (ed.) (2014) *Right to Information: Case Studies on Implementation.* Washington, DC: World Bank Group

Ubaldi B (2013) Open government data: Towards empirical analysis of open government data initiatives. *OECD Working Papers on Public Governance* No. 22. Paris: OECD Publishing

Vickery G (2011) *Review of recent studies on PSI re-use and related market developments.* Paris: Information Economics

Verhulst SG & Young A (eds) (2017) *Open Data in Developing Countries: Toward Building an Evidence Base on What Works and How.* Cape Town: African Minds

Weber M (2009) Politics as a vocation. Translated in: HH Gerth & CW Mills (eds) *From Max Weber: Essays in Sociology.* New York: Oxford University Press. pp. 77–128

Weber M (2015) Bureaucracy. Translated in: T Waters & D Waters (eds) *Weber's Rationalism and Modern Society: New translations on politics, bureaucracy and social stratification.* USA: Palgrave MacMillan. pp. 73–128

World Bank (2003) *World development report 2004: Making services work for poor people.* World Development Report. Washington, DC: World Bank Group

World Bank (2015a) *Tanzania – Open government and public financial management (OGPFM) development policy operation project.* Washington, DC: World Bank Group

World Bank (2015b) Open data for sustainable development. Transport and ICT global practice. Washington, DC: World Bank Group

World Bank (2016) *World development report 2017: Governance and the law.* World Development Report. Washington, DC: World Bank Group

World Wide Web Foundation & Governance Lab (2014) *Towards common methods for assessing OGD: Workshop report and draft framework.* New York: World Wide Web Foundation & GovLab

World Wide Web Foundation (2015) *Open data barometer – Global report* 2nd edn. Washington, DC: World Wide Web Foundation

Yu H & Robinson D (2012) The new ambiguity of open government. *UCLA. Law Review Discourse* 59: 178–208

Zacharia C (2014) *Open government data for effective public participation: Findings of a case study research investigating Kenya's open data initiative in urban slums and rural settlements.* Nairobi: Jesuit Hakimani Centre

About the authors

Michael Cañares is the managing consultant at Step Up Consulting and leads several data for development projects globally, namely, research on open contracting and inclusion in Africa and Asia funded by HIVOS, data-based road asset management capacity building for UNDP in the Philippines, labour market information for ILO in Myanmar, results measurement on road governance projects in Vanuatu and labour market data project feasibility for Indonesia funded by GIZ.

Markus Christian is a researcher at Sinergantara. He holds a master's degree in development studies from Bandung Institute of Technology and has conducted research on basic services and their relation to gender and human rights and other development issues. His particular interest is in understanding the contribution of technology, especially ICTs, to government-people relationships and on how the conception and adoption model is being applied in various ICT initiatives.

Pieter Colpaert is a post-doctoral researcher at Ghent University, IDLab – IMEC. His team researches public web APIs and open data. They also maintain a couple of APIs themselves, such as the Linked Dataset of the 4 public transport operators in Belgium. In 2012, he co-founded Open Knowledge Belgium, an umbrella for organisations sharing the common vision of open knowledge.

Doyinsola Dina, at the time of writing, was the data researcher and analyst of the Open Data Research Centre, School of Media and Communication, Pan-Atlantic University. Her experience in data research and journalism, account planning and strategy, and media planning roles in academic, media and advertising industries where she interacted with data is pivotal to her research prowess. She holds a master's in Media and Communication from the Pan-Atlantic University.

Natalia Domagala leads on data ethics policy at the Cabinet Office, Government Digital Service in the UK. She previously advised on open government and open data policies for the Department for Digital, Culture, Media and Sport in the UK and implemented open data challenges for 360Giving. She has research experience in anthropology, gender, civic tech, and economic development. Natalia received her MSc in Local Economic Development from the London School of Economics and her BA in Anthropology and Media from Goldsmiths University.

Patrick Enaholo holds a doctoral degree in media and communication from the University of Leeds, UK. His research interests include digital/social media and open data with particular focus on their cultural significance in society. He is currently a member of faculty at the Pan-Atlantic University in Lagos, Nigeria, where he also heads the Open Data Research Centre, a research unit focusing on the impact of data on development in developing contexts.

Edson Carlos Germano is a professor and programme coordinator at FIA Business School, Brazil. He has a degree in Mechanical Engineering and a PhD in Business Administration from the University of Sao Paulo (USP), Brazil. With 15 years of experience in the architecture, development and implementation of IT solutions, his research interests include information systems, open government data, information security, public management, open data and its impact on society.

Jack Hardinges is a policy advisor at the Open Data Institute. He works across a number of projects at the ODI to develop, test and communicate policy related to data and technology. Jack joined the ODI in July 2014 and, as a co-author of Open Data Means Business, Open Enterprise, and a number of other reports, he has helped to develop the evidence base for the economic, societal and environmental impact of open data.

Michael Christopher Jelenic is a public sector specialist working in the World Bank's Governance Global, where he has contributed to lending and analytic projects in over 25 countries. His governance work likewise includes researching the impacts of open government, open data, transparency, and public accountability mechanisms as well as providing countries with technical assistance and capacity building support in the implementation of public financial management and public administration reforms.

Therese Karger-Lerchl is a senior economist at Vivid Economics, where she advises private and public sector clients on urban development strategies. When at the Open Data Institute, she focused on using open data to improve public services in cities. Therese previously worked at the South African National Treasury, KPMG and GIZ. She holds an MSc and BSc in Economics, from Barcelona Graduate School of Economics and Maastricht University, respectively.

Bert Marcelis is a system developer working for Samtrafiken in Stockholm, Sweden. He completed his master's in Information Engineering Technology at Ghent University where he was given the TomTom Best engineering thesis of 2018. He also works as a volunteer at the iRail APIs and is an advocate of open data.

Miranda Marcus works for the Open Data Institute. She leads the R&D Programme, an applied interdisciplinary programme combining economics, policy, technology and design to support trustworthy data-driven innovation in the public and private sectors. She is a digital programme manager specialising in agile delivery and digital strategy. Her background is in design and digital anthropology. Her personal research field focuses on the impact of engineering practices on medical artificial intelligence applications.

Julio Paciello is the president and director at Centro de Desarrollo Sostenible in Paraguay. He is a computer engineer with more than 15 years of professional experience in projects of software development, e-government, open data and ICT consultancy. He is also a professor at the National University of Asunción, with many scientific publications.

Juan Pane is the chief innovation officer at Centro de Desarrollo Sostenible in Paraguay. He has a PhD in Computer Science from the University of Trento, Italy and works as a consultant in open data and transparency for the World Bank, USAID, OAS and PAHO in Paraguay. He is also a lecturer and researcher for open data related topics at the National University of Asunción. His work has been published in many scientific papers and journals.

Ed Parkes is Director at Emerging Field, a data strategy and innovation consultancy. Having worked in data strategy in UK central government with organisations such as Ordnance Survey and the Met Office, he now works with social and public sector organisations to help them understand the opportunities to use data. He has worked for the Cabinet Office, Nesta, the Open Data Institute and the Government Digital Service.

Nicolau Reinhard is a senior professor of Management at the School of Economics, Administrations and Accounting of the University of São Paulo (USP), Brazil. His research interests include management of IT function, use of IT in public administration and IT implementation and impacts. Professor Reinhard has a degree in Engineering, a PhD in Management, and has also held executive and consulting positions in IT management in private and public organisations.

Julián Rojas is a PhD researcher at Ghent University, IDLab – IMEC. His work focuses on researching publishing strategies and web interfaces for linked open data. He mainly works on the mobility domain and is currently maintaining the linked open datasets for the main four public transport operators in Belgium. He is an advocate of open data and a main contributor of the open source route planning library Planner.js.

Camila Salazar holds an MSc in Applied Social Data Science from the London School of Economics and Political Science. She has experience as a data analyst in open data initiatives, training and research projects across Latin America and has worked as a data-driven journalist for international media organisations.

Ilham Cendekia Srimarga is a researcher and director at Sinergantara. He holds a master's degree in Development Studies and a bachelor's degree in Informatics from the Bandung Institute of Technology. He has conducted research on issues related to open data, ICT and development, governance, and public policy. He is also a practitioner of ICT and governance, which includes work on developing data revolution strategies in governance and in the social sector.

Violeta Sun holds a bachelor's degree in Business Administration. She received her master's and PhD degrees in Business Administration from the University of São Paulo, Brazil. She is currently a professor in Information Systems at the University of São Paulo. Her main research interests include project management, information technology management, IT infrastructure, IT governance, management systems in health and computer applications in public administration.

Mathias van Compernolle joined the Research Group for Media, Innovation, and Communication Technologies at Ghent University in 2013. He holds a bachelor's degree in social work and a master's degree in Public Administration. He is currently working on his PhD concerning open data and e-government within smart cities. He functions as policy and methodology lead at Smart Flanders. As an open data enthusiast on the board of Open Knowledge Belgium, he keeps track of open data policies in Belgium. Alongside this he lectures politics and organisation studies at Artevelde University College.

François van Schalkwyk is a post-doctoral research fellow at the Centre for Research Evaluation, Science and Technology (CREST), Stellenbosch University. François holds master's degrees in higher education studies (University of the Western Cape) and in publishing studies (Stirling University, UK), and a doctoral degree in science and technology studies (Stellenbosch University, South Africa). François's research areas include critical data studies, higher education studies and scholarly communication.

Roza Vasileva is a PhD candidate in Digital Economy with a focus on open data and cities at the Horizon Centre for Doctoral Training, University of Nottingham, UK. Roza has been an ICT and Innovations Consultant to the World Bank since 2012. She has supported open data and digital government projects in over a dozen countries and conducted Open Data Readiness Assessments (ODRA) following the World Bank standard diagnostic methodology.

Ruben Verborgh is professor of Decentralised Web Technology at IDLab, Ghent University, and a research affiliate at the Decentralized Information Group of CSAIL at MIT. Additionally, he acts as a technology advocate for Inrupt and the Solid ecosystem wherein people and organisations control their own data. He aims to build a more intelligent generation of clients for a decentralised web at the intersection of Linked Data and hypermedia-driven web APIs.

Eveline Vlassenroot holds a bachelor's and a master's degree in Communication Sciences. Having completed additional courses in Information Management & Security, she joined the Research Group for Media, Innovation, and Communication Technologies (www.mict.be) at Ghent University in 2017. She is involved in projects related to e-government and web archiving.

www.ingramcontent.com/pod-product-compliance
Lightning Source LLC
Chambersburg PA
CBHW061242220326
41599CB00028B/5510